New Frontiers on Life Cycle Assessment - Theory and Application

Edited by Antonella Petrillo
and Fabio De Felice

Published in London, United Kingdom

IntechOpen

Supporting open minds since 2005

New Frontiers on Life Cycle Assessment - Theory and Application
http://dx.doi.org/10.5772/intechopen.78248
Edited by Antonella Petrillo and Fabio De Felice

Contributors
Antonella Petrillo, Fabio De Felice, Salvatore Mellino, Iolanda Scudo, Enrique Alberto Huerta-Reynoso, Hector Alfredo López-Aguilar, Jorge Alberto Gómez, María Guadalupe Gómez-Méndez, Antonino Pérez-Hernández, Agata Matarazzo, Fabio Copani, Matteo Leanza Leanza, Aldo Carpitano, Graziano Nicosia, Elen Pacheco, Thiago Santiago Gomes, Genecy Rezende Neto, Ana C. N. Salles, Leila L. Y. Visconte, Ian D. Williams, Simon Kemp, Laurie Wright, Patrick Osborne, Carlos Alberto Mendes Moraes, Magali Rigon, Rafael Zortea, Regina Modolo

First published in London, United Kingdom, 2019 by IntechOpen
IntechOpen is the global imprint of INTECHOPEN LIMITED, registered in England and Wales, registration number: 11086078, The Shard, 25th floor, 32 London Bridge Street
London, SE19SG - United Kingdom
Printed in Croatia

British Library Cataloguing-in-Publication Data
A catalogue record for this book is available from the British Library

Additional hard and PDF copies can be obtained from orders@intechopen.com

New Frontiers on Life Cycle Assessment - Theory and Application
Edited by Antonella Petrillo and Fabio De Felice
p. cm.
Print ISBN 978-1-83880-693-4
Online ISBN 978-1-83880-694-1
eBook (PDF) ISBN 978-1-83880-695-8

For EU product safety concerns:
IN TECH d.o.o., Prolaz Marije Krucifikse Kozulić 3, 51000 Rijeka, Croatia, info@intechopen.com or visit our website at intechopen.com.

Meet the editors

Antonella Petrillo is a professor at the Department of Engineering of the University of Naples "Parthenope," Italy. She received her PhD in Mechanical Engineering from the University of Cassino. Her research interests include sustainability, multicriteria decision analysis, industrial plant, manufacturing, and safety. She serves as an associate editor for the International Journal of the Analytic Hierarchy Process. She is a member of the AHP Academy and a member of several editorial boards.

Fabio De Felice is a professor at the University of Cassino and Southern Lazio, Italy, where he received his PhD in Mechanical Engineering. His current research focuses on multicriteria decision-making analysis (with an emphasis on AHP and ANP) and industrial, project, and supply chain management. Currently, he serves as a member of the Scientific Advisory Committee of the International Symposium on the Analytic Hierarchy Process. He is the founder of the AHP Academy that promotes the diffusion of culture and methodologies of decision-making, with particular reference to those based on AHP. He is a member of the editorial boards of several international organizations and journals and has authored/coauthored numerous articles in the areas of decision science and business management.

Contents

Preface

In recent years, the topic of life cycle management has become a challenge for the industrial and service sectors. There is a growing understanding that businesses cannot focus only on short-term profitability.

At the same time, process optimization and efficiency are profoundly characterizing the industrial sector, requiring the application of new algorithmic and theoretical techniques.

The deep integration of the circular economy with industrial systems is also changing the relationship between products, processes, and society.

In this regard, life cycle assessment (LCA) is a "holistic approach" that supports decision-makers in various decisions such as the selection of processes, materials, and supply chains.

The book contains original research and application chapters from different perspectives. It is enriched through the analysis of case studies.

In detail, Chapter 1 proposes an LCA study to quantify which components of a bicycle have the highest environmental impact. The use of two different bicycles in Italy, an electric pedal-assisted bike and a hydrogen fuel cell-operated bike, is analyzed. Chapter 2 presents a generalized model for the construction of inventories regarding the production of electricity through biogas. The LCA of two types of plants is presented as case studies: (1) the generation of electrical energy from organic waste in sanitary landfills and (2) the generation of electric energy from dairy cattle manure. Chapter 3 analyses the wine production chain according to the preliminary LCA model. LCA has been implemented on the whole production chain of the product "Lenza di Munti," a bottle of wine by "Nicosia S.p.a." The chapter aims to provide a complete picture of the interactions between the product and the environment, to understand the environmental consequences and provide the necessary information to define the best solutions. LCA for waste tires is presented in Chapter 4. Tires are complex materials manufactured from vulcanized rubber and various other reinforcing materials. LCA has been used to quantify their impact and support the decision-making process to determine the most beneficial alternative from an environmental standpoint. Perspectives on subnational carbon and climate footprints, a case study of Southampton, UK, are analyzed in Chapter 5. This chapter develops the data and methods required for subnational territorial, transboundary, and consumption-based carbon and climate footprints. The results and implications of each footprinting perspective are discussed in the context of emerging international standards. The study clearly shows that the carbon footprint (CO_2 and CH_4 only) offers a low-cost, low-data, universal metric of anthropogenic greenhouse gas emissions and subsequent management. Finally, Chapter 6 focuses on the topic of renewing a university to be able to support the adaptation of Industry 4.0 within a region. The chapter introduces the main LCIA methods used and the most relevant categories of environmental impact. In total, 87 articles were initially retrieved using relevant keywords.

This book is intended to be a useful resource for anyone who deals with environmental and sustainability problems. Furthermore, we hope that this book will provide useful resources, ideas, techniques, and methods for further research on these issues.

As editors of this book, we very much thank the authors who accepted to contribute with their invaluable research, as well as the referees who reviewed these papers for their effort, time, and invaluable suggestions. Our special thanks go to Ms. Marina Dusevic, Author Service Manager, for her precious support and her team for this opportunity to serve as guest editors.

Antonella Petrillo
Professor,
Department of Engineering,
University Parthenope of Naples,
Naples (NA), Italy

Fabio De Felice
Professor,
Department of Civil and Mechanical Engineering,
University of Cassino and Southern Lazio,
Cassino (FR), Italy

Chapter 1

Design of a Sustainable Electric Pedal-Assisted Bike: A Life Cycle Assessment Application in Italy

Antonella Petrillo, Salvatore Mellino, Fabio De Felice and Iolanda Scudo

Abstract

Transport is one of the economic sectors within the European Union with the most detrimental effects on climate change. In this context, electric bicycles (e-bikes) are considered as a potentially effective technological innovation to reduce carbon impacts. The present research aims to propose a life cycle assessment study to quantify which components of a bicycle have the highest environmental impact. The use in Italy of two different bicycles, an electric pedal-assisted bike and a hydrogen-fuel cell-operated one, is analyzed. The final aim of the study is to quantify and to evaluate the bike's energetic and environmental performances, focusing the analysis on the vehicle production and use phases. To achieve the abovementioned purpose, two approaches, the "from cradle to grave" approach and "well to wheel" approach, are considered.

Keywords: sustainability, smart mobility, LCA, electric pedal-assisted bike, hydrogen-fuel cell bike

1. Introduction

Transport is one of the main causes of air pollution, climate change, and urban noise problem [1]. In fact, it is responsible for about one-third of the global consumption of energy and more than one-fifth of the damage caused by the greenhouse effect, global warming, CO_2 emissions, and human health [2]. In recent years, policies and incentive campaigns have been promoted for electric mobility in order to reduce the use of fossil resources and to limit the use of polluting motor vehicles.

The aim is to promote sustainable mobility, preserving the environment and improving the quality of life. Modern society is increasingly moving toward the use of zero-impact transport systems able to satisfy the needs of the user while achieving a reduction in consumption and harmful emissions [3]. The academic community and producers are paying close attention to a particular type of electric mobility that is represented by the *electric pedal-assisted bikes*. Electric pedal-assisted bikes have conquered the electric vehicle market, positioning themselves at the top of the classification, thanks to the technological innovation and to the cost reduction of electronic components [4]. Also known with the term *pedelec*, in the last decade, they have become one of the most used means of transport in the city thanks to the presence of the electric motor to support the pedaling [5].

The sector of electric bikes underwent strong development from 2006 to 2016 [6]. According to recent statistics, conducted in Italy, Germany, France, the United Kingdom, Holland, and Austria by Claus Fleischer, CEO of Bosch eBike Systems, it emerged that 10% of interviewed is already an e-bike owner, while 16% are considering buying one within the following year; preferences are oriented toward e-bike by city (29%), followed by trekking models (11%), urban (9%), and e-mountain bikes (8%), thus expecting an even more exponential growth for the next decade [7].

These ecological vehicles have their specific characteristics: (1) electric bicycles are not mopeds and (2) their speed is modest. This feature makes electric bikes usable by anyone, even without a license and without wearing a helmet.

These vehicles make it possible to combine the advantages of traditional bikes with those of mopeds, so much so as to be called the "ecological bicycles of the planet." The tendency is confirmed by the fact that the fight against pollution has pushed many cities of the world to rethink their mobility by favoring zero-impact vehicles.

The analysis of the state of the art of life cycle assessment (LCA) studies applied to the electric pedal-assisted bikes showed that studies on pedelec are still very few [8]. An interesting study was conducted by Blondel et al. [9] in November 2011 on behalf of the Financial European Commission, where a complete LCA analysis was carried out by the different types of transport (bicycle, pedelec, bus, car) in order to evaluate the CO_2 emissions from each of these vehicles during production, use, and disposal.

The aim of the present research was to evaluate the energy-environmental performance of an electric pedal-assisted bike, using the SimaPro© software. Specifically, the analyzed pedelec is the e-bike produced by an *Italian artisan company located in Milan (North Italy)*, active in the two-wheeler sector since 1908. The study followed the approaches "from cradle to grave" and "well to wheel."

The rest of the chapter is organized as follows: Section 2 presents a brief history of electric bikes; Section 3 analyzes the main rules and regulations governing electric bicycles; in Section 4 systems under study are presented; in Section 5 the methodological approach and the main assumptions of the study are explained; and, finally, Section 6 summarizes the main results and implications of the study.

2. A brief history of electric bikes

Normally, it is thought that e-bike is a relatively recent invention. The history of the ebike was made by brilliant and passionate inventors. The first bicycles with electric motors appear at the end of 19th century [10]. More precisely in 1895, Ogden Bolton Jr., an American inventor, decided to apply for the first time an electric motor on the rear hub of a traditional bicycle [11]. For many years, due to reduced autonomy, the production of electric bikes was difficult; electric bikes remained as simple prototypes. Only in the 1930s, more and more European companies began to produce complete models of e-bikes, even if the currently available technology did not yet give the possibility of producing light vehicles with great autonomy as happens today. One of the first successful models was made in 1937 by Philips, a famous and active company in the radio sector, in partnership with Gazelle, a company that produces traditional bikes of great renown in the Netherlands and in Europe (**Figure 1**). The model, which made 117 sales, had a weight of 50 kg net of the battery that was characterized by a range of 40 km. It needs to do a full day to charge the battery. The vehicle reached a speed of about 18 km/h.

Later in 1946, Benjamin Bowden, an Anglo-American industrialist very active in the automobile sector, launched the Spacelander, known as "The Bike of the Future" (**Figure 2**).

The bike, characterized by a modern design, was very interesting from a technical point of view, introducing ideas still valid today. Both the battery and the cables

Figure 1.
The Philips-Gazelle 1937 electric bike (source: bikeitalia.it).

Figure 2.
The Spacelander (source: bikeitalia.it).

were integrated into the chassis, while the engine positioned in the rear hub also functioned as a dynamo to recharge the battery. Furthermore, according to the original patent, in a 10% climb, it could reach a speed of 8 km/h. However, the bike that represented a very high-end product, evidently too advanced for the time, was never produced on a large scale.

It is worth mentioning that from the second half of the twentieth century the low ecological conscience and cheap oil encouraged mass motorization reducing the interest in traditional and electric bikes. Only later, in the 1970s, the enormous rise in the price of crude oil and the spread of the first ecological movements renewed interest in electric vehicles [12]. The years following 1973 were of extreme importance for the technology behind electric bicycles. More and more designers began to patent e-bike models. In 1990, Michael Kutter developed what some people feel is the first Pedelec (PEDal-ELECtric). It later came to be referred to as Pedal Assist System (PAS). The first production models were sold in 1992 under the Dolphin name for the Swiss company Velocity, but they did not survive. In 1993 Yamaha developed its first electric bicycle, which included the now popular Pedal-Assist-System (PAS). Similarly, in the same period, other companies started their productions in Europe and Italy.

Subsequently, thanks to the use of sealed lead batteries and the experimentation of the lightest nickel-cadmium, electric bicycles were produced which are very similar to those of today, on a large scale and with affordable prices.

In the last 20 years, the electric bike models available to the public have multiplied dramatically, and several manufacturers have experimented with innovative technical solutions to achieve a better integration between pedaling and engine assistance. Today, the technological innovations have made the e-bike a vehicle to all effects capable of optimally replacing a car over medium-short distances.

3. Rules and regulations governing electric bicycles

To be considered as "lawful" bicycles and to be treated as traditional bicycles, electric bicycle models must possess the famous "Pas," i.e., the Pedal Assist Systems, facilitated by the electric motor [13]. There are two *macro categories* of electric bicycles: (1) *electric bicycles* (or e-bike) with acceleration system and (2) *electric pedal-assisted bicycles* (or pedelec).

The *electric bicycle* that owns the acceleration system is equipped with an electric motor driven by an accelerator lever, generally positioned on the handlebar that allows you to accelerate independently of the use of pedals. This type of ecological vehicle is not really an electric bike but is categorized as an "electric scooter," and as such, it is the object of matriculation, license plate, insurance, wearing a helmet, and all the operations to be carried out in the case of motor vehicles.

The *electric pedal-assisted bicycle*, on the other hand, is equipped with an intelligent electronic control unit that understands when the driver is using the pedals and when not activates the electric motor only if the pedals are in motion.

The European Directive 2002/24/EC specifies the characteristics of pedelecs and defines the pedal-assisted cycles as "bicycles equipped with an electric auxiliary motor having a maximum continuous nominal power of 0.25 kW, whose power is progressively reduced and subsequently interrupted when the vehicle reaches 25 km/h, or sooner if the cyclist stops pedaling" [14].

This definition is included in the Italian regulations or in Article 50/2009 of the Highway Code. The definition does not limit to this value the maximum speed of the bike but indicates that it can be overcome by relying only on the capabilities of the user, not being the pedelec equipped with an acceleration system.

The European Standard EN 14764 standard "City and trekking bicycles: Safety requirements and test methods," specifies safety and performance requirements for the design, assembly, and testing of bicycles and sub-assemblies intended for use on public roads, and lays down guide lines for instructions on the use and care of such bicycles.

In addition, the Directive 2002/24/EC of the European Parliament and of the Council of 18 March 2002 relating to the type-approval of two or three-wheel motor vehicles and repealing Council Directive 92/61/EEC establishes that all two or three-wheel motor vehicles must be equipped with the CE conformity marking consisting of the homonymous "CE" symbol.

In addition, each bicycle in accordance with the law must be accompanied by an instruction manual useful to the purchaser, containing the main information for the correct use of the vehicle and for its maintenance.

4. Systems description

The product analyzed in this study is not a generic pedelec is not a generic pedelec, but rather an electric pedal-assisted bike. Specifically, the prototype studied is the classic Corinto model produced by Italian artisan company located in Milan (Northern Italy), to which the bike + all-in-one system of Zehus human+ has been incorporated, as shown in **Figure 3**.

The prototype has the following features:

- Aluminum frame
- Bike + all-in-one propulsion system from Zehus human + with:

Figure 3.
The pedelec prototype.

- ○ Brushless motor, with rated power of 250 W and nominal voltage of 24 V, complete with engine control unit (ECU)
- ○ Lithium battery with battery management system (BMS) associated with the capacity of 160 Wh
- ○ A control unit (pedaling sensor—PAS, two speed microcontrollers, integrated Bluetooth module for controlling the vehicle)

- Rear and front rod brakes
- 28-inch rubber wheels
- Saddle and handlebar from the Miele B66 series, made by Brooks

The battery, recharged by a simple power outlet, powers the engine, which is started as soon as the pedal sensor detects the movement of the pedals. The whole system, thanks to the presence of the control unit and the Bluetooth unit, is controlled using a telephone app for Android and iOS systems.

The characteristics of the H-bike considered as well as the data refer to the articles of Kheirandish et al. [15], Cardinali et al. [16] and Mellino et al. [17] and are summarized in the following. The H-bike has a 250 W PEM fuel cell system, an electronic control unit (ECU), a rechargeable nickel-metal hydride (Ni-MH) battery of 25 V and 9 Ah, 2 converters 150W DC/DC, a 150 W electric motor and a hydrogen storage tank. The PEM cell, consisting of 3 main parts: cathode, anode and proton exchange polymer membrane (Nafion), transforms the chemical energy, released during the electrochemical reaction between hydrogen and oxygen, into electrical energy.

5. Materials and methods

5.1 Goal and scope definition

The aim of the study was to evaluate the energy-environmental performance of a pedal-assisted bicycle prototype, according to the ISO 14040 series standards and the International Reference Life Cycle Data System (ILCD) [18, 19].

The study, conducted according to the life cycle assessment (LCA), is focused on the analysis of the *production* and *use* phases of the vehicle. Specifically, the phases of the life cycle were examined, involving the extraction of raw materials up to the finished product and from the finished product to use. A comparative analysis was carried out between the pedelec prototype and the H-bike, an equally innovative technology. The analysis was carried out considering two main approaches one for each phase:

Phase #1: "From cradle to grave" approach. An assessment was carried out of the environmental impacts of the production of vehicles and of the elementary components starting from the extraction, production, and use, up to the recycling and disposal of the residual waste.

- *Phase #2: "Well to wheel" approach.* The environmental impacts and energy vectors used during pedelec prototype operation have been estimated. Materials and the energy used have been evaluated according to a "well to wheel" approach, i.e., considering the phases of production of energy carriers (electricity and hydrogen).

In the first case, a bicycle unit was chosen as the functional unit, to which all materials, emissions, and energy consumption were referred.

Instead, in the second case, it was used as a functional unit 100 km.

Really, transport studies use a functional unit expressed in passengers for kilometers traveled p-km. However, in our study, being the pedelec prototype used by a single person it was considered appropriate to refer to the distance only. Nevertheless, in the **Figure 4** the scheme used for the assessment of the usage phase is given.

5.2 Main assumptions

Once the functional units have been defined, some assumptions have been made to conduct the study, as defined below:

Hypothesis 1: The same bicycle structure (for both bikes) was considered in terms of size, composition, and materials to concentrate the study on electrical technologies and evaluate the real environmental impacts of the two different systems.

Hypothesis 2: For the pedelec prototype, which has a lithium battery of 30 V and 5.3 Ah, 1000 recharging cycles have been considered, for which an efficiency of the processes of charge and discharge constant throughout the life span has been

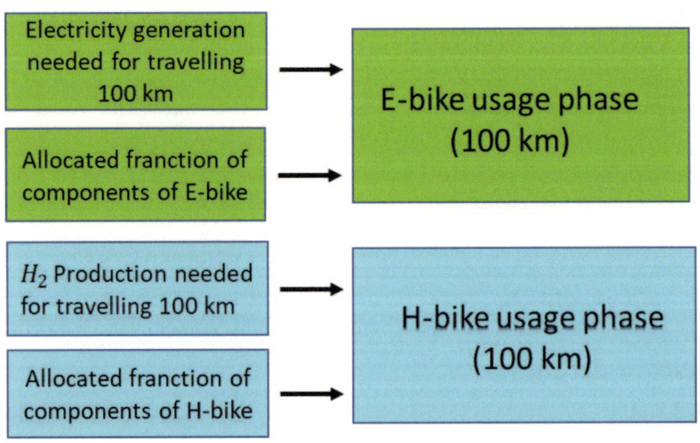

Figure 4.
LCA scheme used for assessing the usage phase.

assumed. Moreover, during use, the speed of the vehicle was set at 25 km/h. It was considered that compared to a full charge the autonomy is 35 km with activation of the kinetic energy recovery system (KERS) system and 25 km in case of absence energy recovery braking.

Hypothesis 3: According to *Hipotesis 2*, a life span of 35,000 km was considered in the first case and 25,000 km in the second.

Hypothesis 4: For H-bike, on the other hand, a duration of the PEM fuel cell equal to 120,000 km was hypothesized, as suggested by the US Department of Energy (https://www1.eere.energy.gov/hydrogenandfuelcells/pdfs/accomplishments.pdf).

Hypothesis 5: Subsequently, according to hypothesis 1, for both cases, a life span of the bicycle and the other system components of 120,000 km was assumed.

Hypothesis 6: Furthermore, it was assumed that the hydrogen, used during the use phase, was produced by electrolysis from a photovoltaic refueling station; while for recharging the lithium battery, electricity taken directly from the network was used first, then from the typical Italian energy mix, and then subsequently, the electricity withdrawn from the grid, produced through a renewable electric mix. For this purpose, the mix proposed by the Enel Green Power society was used (https://www.enelgreenpower.com).

5.3 Inventory analysis

In the inventory analysis phase, the input and output flows of material and energy were identified and quantified, and the emissions into the air, water, and soil involved in the production phase of the e-bike. The data were provided in part by the company (see **Tables 1** and **2**) and partly retrieved from the literature and from the databases contained in SimaPro©. Ecoinvent database was used since it is one of the most complete and the most used for European LCAs [20].

In particular, for the structural part of the bike, a process already existing in SimaPro© was used with the name "bicycle," which was appropriately modified according to the information obtained and the values identified in order to adapt it to the prototype.

For the propulsion system, it was considered an electric motor of a scooter, which for size was close to that of Zchus. While, the control unit has been schematized with a generic process called "electronics, for control unit", having no detailed information on the sensors and the materials used.

Instead, for the lithium battery, we used the "battery cell, Li-on" process that reproduces it faithfully, modifying the data on the battery capacity and inserting the weight of the battery. **Figure 5** shows the process tree for production phase of pedelec prototype.

Components	Materials	Length	Outer diameter	Internal diameter
Horizontal pipe	Aluminum	595 mm	36 mm	32 mm
Vertical tube	Aluminum	470 mm	36 mm	32 mm
Oblique tube	Aluminum	520 mm	36 mm	32 mm
Horizontal sheath	Aluminum	420 mm	25 mm	—
Vertical sheath	Aluminum	435 mm	25 mm	—
Steering tube	Aluminum	140 mm	36 mm	32 mm
Fork	Aluminum	524 mm	30 mm	—

Table 1.
Pedelec prototype frame data (source: Italian artisan company located in Milan (North Italy)).

Components	Technical features	Weight
Engine	250 W–24 V	1.7 kg
Battery	160 Wh–5.3 Ah	600 g
Control unit	Four sensors	200 g
Coating	Aluminum	500 g

Table 2.
Bike + all-in-one propulsion system data (source: Zehus).

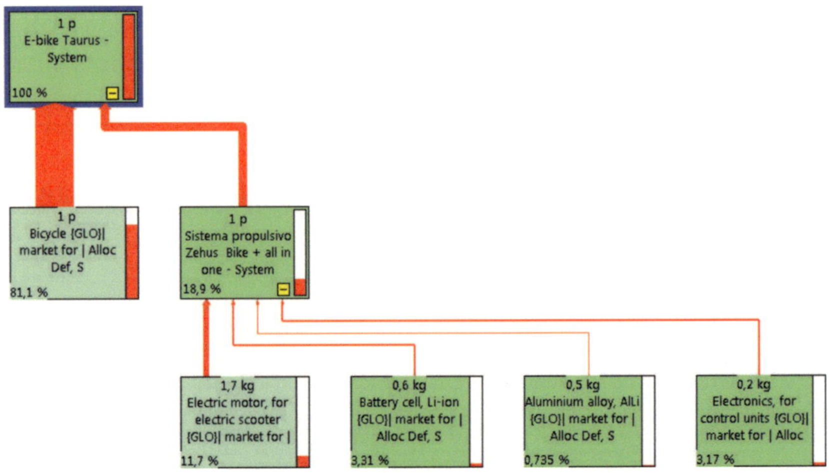

Figure 5.
Process tree for production phase of pedelec prototype (SimaPro©).

Instead, with reference to the H-bike, used as a comparison system, the data necessary for the reconstruction of the model were taken from the study conducted by Mellino et al. [17]. Specifically, for the bike structure, the same e-bike process was used, to analyze the two models with the same chassis, while for the innovative technology, all the processes necessary for the reproduction of the PEM cell were recreated. In the hydrogen bike model, have been added the following processes:

1. The "bicycle" and "PEM stack cell" processes to the hydrogen accumulation;

2. The process for the nickel-metal hydride storage battery

3. The processes "electric motor, for electric schooter" and "convert, for electric passenger car" to recreate the engine and the DC/AC converter. The process tree is shown in **Figure 6**.

Instead, for the use phase of the e-bike, not yet this commercialized, reference was made to data in the literature and in the Zehus manual. In fact, 1000 cycles of recharging have been considered for the lithium battery, and the speed of the bike has been set at 25 km/h. Thus, the manual has been identified as the autonomy of the bike in terms of km traveled equal to 35 km with activation of the KERS and 25 km in the absence of energy recovery during braking. Depending on the cycles and speed, and therefore of the kilometers traveled, the battery wear was defined not in terms of cycles but in terms of km having to relate to the functional unit (100 km). In fact, it was equal to 35,000 km in the first case and 25,000 km in the

DOI: http://dx.doi.org/10.5772/intechopen.81737

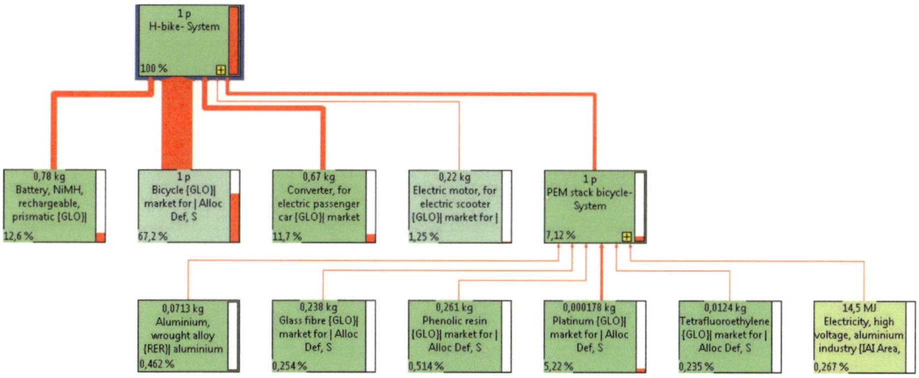

Figure 6.
Process tree for production phase of H-bike (SimaPro©).

second. Based on this information, a complete recharge of 160 Wh is required; the use phase reconstructed in the following figures has been reconstructed.

For the use phase of the H-bike, the data were retrieved only from the literature. A study conducted by the Department of Energy of the University of the United States has identified the data on the wear of the PEM cell equal to 120,000 km. The study conducted by Mellino et al. [17] has found the amount of hydrogen equal at 75 g, to complete 100 km. For which the information is known similarly before the tree has been recreated for the use phase of the H-bike.

Similarly, process trees for use phase were performed with SimaPro©.

5.4 Results from the Impact Assessment

The energy-environmental performance of the prototype was assessed in the production and use phases of the vehicle, using the IMPACT 2002+ method [21].

5.4.1 Classification

Impact categories have been chosen according to the Impact 2002+ method to have a global vision of the impacts of the pedelec in the production and use phases. In all, 14 impact categories were analyzed. The considered midpoints are shown together with the respective reference substances in **Table 3**.

5.4.2 Characterization

Using characterization factors typical of the IMPACT method, the scores of the different elements according to the kilogram equivalents of the reference substance of each category were quantified for each impact category. **Figure 7** shows the environmental impacts of pedelec and H-Bike for each impact category for production phase.

Similarly, we calculated the impacts for the use phase.

Figure 8 shows the results obtained from the comparison of the e-bike, in the case with and without KERS activation, with the hydrogen bike.

For the use phase, the process was also recreated assuming that the energy used to recharge the battery is produced from renewable sources through the "Green Mix of Enel" project. **Figure 9** shows environmental impacts of pedelec, with and without KERS, and H-Bike for each impact category for use phase and in the case of Enel Green Power.

Midpoint	Units
Carcinogens	kg C2H3Cl eq
Non-carcinogens	kg C2H3Cl eq
Respiratory inorganics	kg PM2.5 eq
Ionizing radiation	Bq C-14 eq
Ozone layer depletion	kg CFC-11 eq
Respiratory organics	kg C2H4 eq
Aquatic ecotoxicity	kg TEG water
Terrestrial ecotoxicity	kg TEG soil
Terrestrial acid/nutri	kg SO2 eq
Land occupation	m2org.arable
Aquatic acidification	kg SO2 eq
Aquatic eutrophication	kg PO4 P-lim
Global warming	kg CO2 eq
Nonrenewable energy	MJ primary
Mineral extraction	MJ surplus

Table 3.
Impact and damage categories; IMPACT 2002+ method.

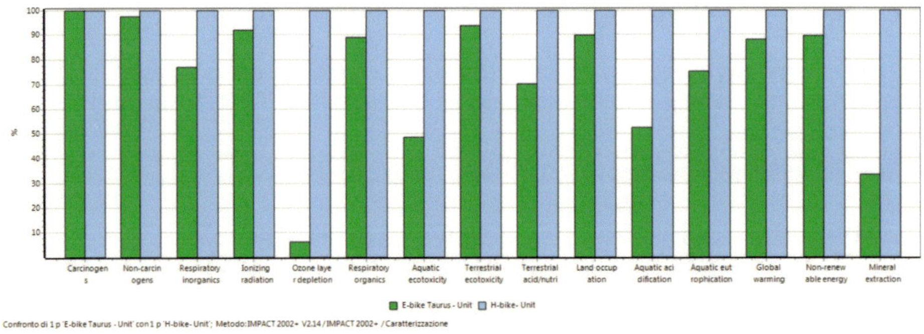

Figure 7.
Environmental impacts of pedelec and H-Bike for each impact category for production phase.

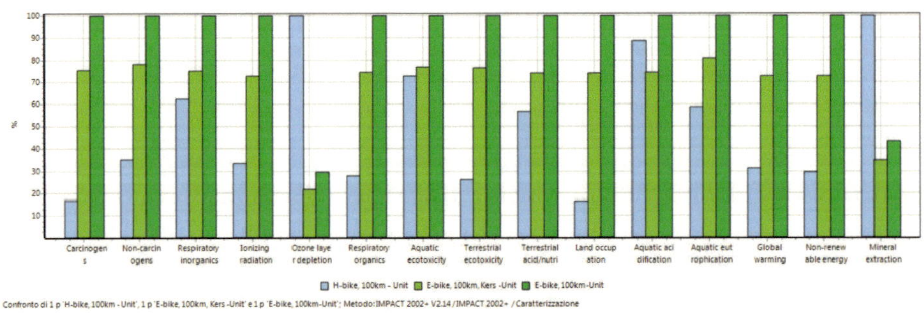

Figure 8.
Environmental impacts of e-bike, with and without KERS, and H-Bike for each impact category for use phase.

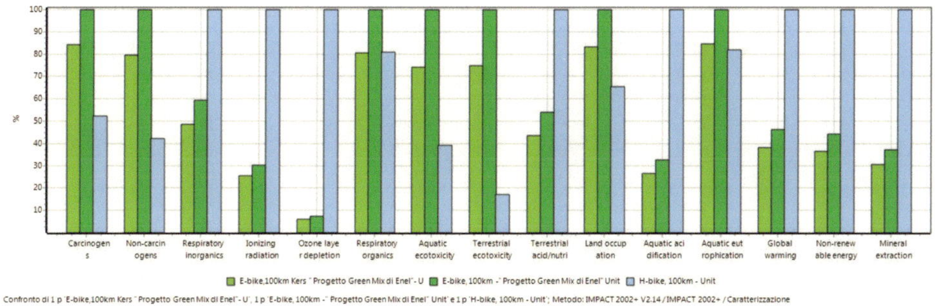

Figure 9.
Environmental impacts of e-bike, with and without KERS, and H-Bike for each impact category for use phase (Enel Green Power).

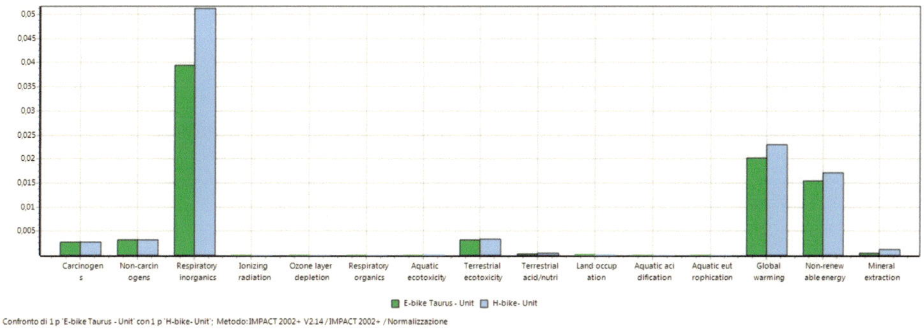

Figure 10.
Comparison of the impact categories between e-bike and H-bike for production phase.

5.4.3 Normalization

This phase aims to determine the importance of the individual environmental effects, dividing the impacts of each category by the respective reference values, represented by average data in a world, regional, or European scale, referring to a given time interval, generally 1 year.

Figure 10 shows comparison of the impact categories between pedelec and H-bike, for production phase.

The results show that of the 14 impact categories the ones that most weigh on the final pollution are essentially the following three:

- Respiratory inorganics
- Global warming
- Nonrenewable energy

Similarly, the procedure was subsequently applied to the use phase.

A greater impact of the categories has been confirmed: respiratory inorganics, global warming, and nonrenewable energy.

5.4.4 Damage assessment

In this phase, the impacts, identified in the characterization, were grouped into four categories of damage, human health, ecosystem quality, climate change, and resources, according to the logic shown in **Table 4**.

Impact categories	Damage factors	Damage categories	Unit of measurement
Carcinogens	2,80 E-6	Human health	DALY
Non-carcinogens			
Respiratory inorganics	7,00 E-4		
Ionizing radiation	2,10 E-10		
Ozone layer depletion	1,05 E-3		
Respiratory organics	2,13 E-6		
Aquatic ecotoxicity	5,02 E-5	Ecosystem quality	PDF*m2*yr
Terrestrial ecotoxicity	7,91 E-3		
Terrestrial acid/nutri	1,04		
Land occupation	1		
Aquatic acidification	8,86 E-5		
Aquatic eutrophication	8,86 E-5		
Global warming	1	Climate change	*kg di* CO_2
Non-renewable energy	1	Resources	MJ
Mineral extraction	1		

Table 4.
Damage categories.

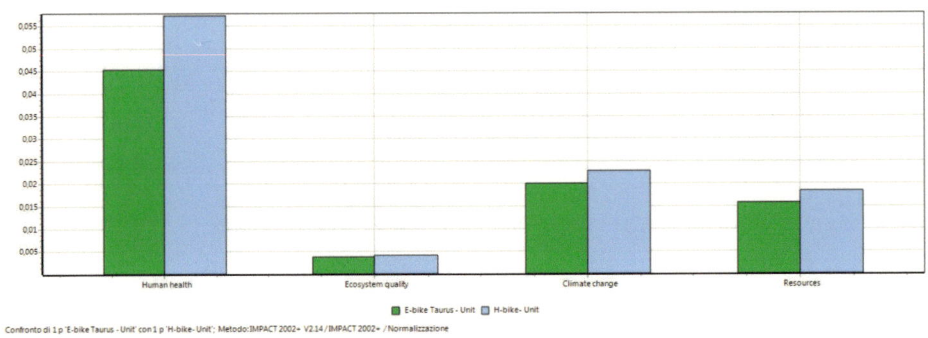

Figure 11.
Comparison of the damage categories of pedelec with H-bike for production phase (normalized).

Figure 11 shows the four endpoints identified for the production of the pedelec. The values identified were then normalized.

Similarly, the damage categories for the use phase of the pedelec, with and without KERS (with and without normalization), were identified.

The comparison of the impact categories of the pedelec has also been calculated with and without activation of the braking energy recovery system with those of the H bike, again in the use phase.

5.5 Discussion

5.5.1 Comparative analysis of the production phase

The results show that the processing of the e-bike is less impactful in all the categories considered than that of the H-bike.

This result is due to a greater complexity of the H-bike, which requires the use of special materials and complex processes for the control unit and the PEM fuel cell along the production chain.

Both processes mainly impact on the respiratory inorganics category, even if there are not negligible effects on the categories of global warming and nonrenewable energy, while in the others, there are minor impacts, almost negligible.

As far as respiratory inorganics is concerned, the pedelec generates a 1.2 times lower impact than the H-bike (respectively, 0.4 and 0.519 kg PM 2.5 eq), thanks to the presence of a structurally simpler propulsion system than to what characterizes the hydrogen bike.

Specifically, analyzing the different contributions of the processes related to the construction of the pedelec, it is clear that the greatest impacts are generated by the structure of the bike, due to the processing of large quantities of aluminum.

As the objective of the present research was to assess the sustainability of electrical technology, the impact categories for the bike + all-in-one propulsion system were analyzed. It emerges that the major impacts are caused by the electric motor of the Zehus system, which accounts for 61.31%, followed by the control electronics and the lithium battery, with the respective percentages of 17.46 and 17.38%.

Regarding the impact of the electric motor, the cause is closely linked to the use of copper for the construction of the rotor, as well as the use of high- and medium-voltage energy, produced from nonrenewable source, used during the process of engine production.

For the control unit, the impact is associated with the main body of the control unit, i.e., the board, which requires rather complex operations for the realization of the components that constitute it, as well as the use of not negligible energy, for which it has been supposedly a world energy mix, considering that most of the electric elements are not manufactured in Italy, but rather imported.

For the lithium battery, the greatest impact is related to the production processes carried out for the construction of anode and cathode.

For the global warming impact category, it is still the engine that generates the greatest impact with 60%, even if this time the cause is no longer made of copper but rather the harmful substances emitted during the processing of permanent magnets used in the brushless motor and to an equal extent by the production processes of aluminum and by the use of medium-voltage energy always produced from fossil fuels, followed by the control electronics and lithium battery whose cause is always represented by aluminum.

Despite this, the electric bike is still better than the H-bike, whose high impacts are linked to the nickel metal hydride accumulation battery, to the processing of platinum used in the construction of the cell and to the complex control unit. In fact, unlike the one installed on the e-bike, it is responsible for supervising the system parameters such as stack temperature, hydrogen pressure, cell voltage and current, battery voltage, and start-up and shutdown of the system, to control the environmental conditions that influence the behavior of PEM.

5.5.2 Comparative analysis of the use phase

The results show that unlike what is seen for production, it is the H-bike that impacts less in all the midpoints considered, except for ozone layer depletion and mineral extraction.

The pedelec bike with or without KERS causes respiratory categories inorganics, global warming, and nonrenewable energy the greater pollution than the hydrogen bike.

More specifically, in the case of activation of the energy recovery system, it generates a lower impact than the bike without KERS but 3.2 times higher than the H-bike in the respiratory inorganics, 2.34 times in global warming, and 2.46 times in nonrenewable energy.

In the absence of the braking energy recovery system, the impacts are even higher.

Renewable source	Quantity
Hydroelectric	49.90%
Geothermal	23.80%
Wind	23.70%
Solar	2.60%

Table 5.
Enel Green Mix project data (source: Enel Green Power).

The pollution, for all three midpoints, is closely linked to two causes:

1. The battery recharge process, and

2. The use of the Italian electric mix produced for the most part (a nonrenewable source).

In fact, in the case of H-bike, the energy used to produce hydrogen is generated by solar radiation.

Since energy is the main cause of pollution, the use phase has been repeated assuming the use of renewable energy, in order to evaluate the dependence on the performance of the e-bike from the Italian electric mix. In fact, a green and renewable production scenario was built, starting from the data of the "Green Mix" project by Enel Green Power reported in **Table 5**.

The impacts of the pedelec have been reduced compared to the phase previously shown. In addition, in the categories where it first had the most impact on the pedelec, the impacts decreased compared to the H-bike. This demonstrated the sustainability of the vehicle and the strong dependence of the pedelec performance of the electric mix used.

6. Conclusions

In the present research, the life cycle assessment methodology was used to evaluate the energy-environmental assessment of the pedelec prototype during the production and use phases of the vehicle. The results identified were compared with the results obtained in H-bike. The main consideration of the study can be summarized as follows:

- In the production phase (developed according to the "from cradle to grave" approach) emerged that the construction of the H-bike has more impact than the pedelec prototype in all the midpoints considered due to the presence of a rather complex control unit.

- The most affected category was respiratory inorganics, followed by global warming and nonrenewable energy. The causes are closely linked to the processing of aluminum and copper, to the harmful substances emitted and to the consumption of energy from the Italian and global electric mix, for the construction of the bike structure and the electric propulsion system.

- The environmental performance of the pedelec deteriorates compared to those of the H-bike when the boundary is moved to the operational phases of the vehicles, due to the use of the Italian electric mix for the battery charging process.

- The environmental performance of the pedelec bike in the use phase improves compared to those of the H-bike when the energy used to recharge the lithium battery is produced from renewable sources.

- The energy mix used for the production phase and for the transport phase influences the performance of the vehicles, generating not negligible environmental impacts.

Considering these results, future efforts are still needed to further reduce the impacts of the pedelec; improvements can be made on the materials used to build the pedelec structure and the propulsion system.

The aluminum, copper, and steel components could be replaced with recycled or more innovative materials. However, electrical technology is still a valid and clean solution for the world of transport, even if now the costs remain quite high.

In addition, future research aims to investigate the disposal and recycling phases.

Conflict of interest

The authors certify that they have no conflict of interest.

Author details

Antonella Petrillo[1*], Salvatore Mellino[2], Fabio De Felice[3] and Iolanda Scudo[1]

1 Department of Engineering, University of Napoli "Parthenope", Napoli, Italy

2 Department of Science and Technology, University of Napoli "Parthenope", Napoli, Italy

3 Department of Civil and Mechanical Engineering, University of Cassino and Southern Lazio, Cassino, Italy

*Address all correspondence to: antonella.petrillo@uniparthenope.it

References

[1] Burgess M, King N, Harris M, Lewis E. Electric vehicle drivers' reported interactions with the public: Driving stereotype change? Transportation Research Part F: Traffic. 2013;**17**:33-44

[2] Pierce JMT, Nash AB, Clouter CA. The in-use annual energy and carbon saving by switching from a car to an electric bicycle in an urban UK general medical practice: the implication for NHS commuters. Environment, Development and Sustainability. 2013;**15**(6):1645-1651

[3] Todorovic M, Simic M. Current state of the transition to electrical vehicles. Smart Innovation, Systems and Technologies. 2019;**98**:130-139

[4] Behrendt F. Why cycling matters for electric mobility: Towards diverse, active and sustainable e-mobilities. Mobilities. 2018;**13**(1):64-80

[5] Munkácsy A, Monzón A. Impacts of smart configuration in pedelec-sharing: Evidence from a panel survey in Madrid. Journal of Advanced Transportation. 2017;**2017**:4720627. 11 pages

[6] Abagnale C, Cardone M, Iodice P, Marialto R, Strano S, Terzo M, et al. Design and development of an innovative e-bike. Energy Procedia. 2016;**101**:774-781

[7] European Bicycle Market. Confederation of the European Bicycle Industry. Brussels, Belgium; 2017

[8] Lee J-S, Jiang J-W, Sun Y-H. Design and simulation of control systems for electric-assist bikes. In: Proceedings of the 2016 IEEE 11th Conference on Industrial Electronics and Applications, ICIEA 2016. Piscataway, NJ: IEEE Operations Center; 2016. pp. 1736-1740

[9] Benoît B, Mispelon C, Ferguson J. Cycle more often 2 Cool down the planet! Quantifying Co2 savings of cycling. Brussels, Belgium: European Cyclists' Federation aSBl; 2011

[10] Ji S, Cherry CR, Bechle MJ, Wu Y, Marshall JD. Electric vehicles in China: Emissions and health impacts. Environmental Science and Technology. 2012;**46**(4):2018-2024

[11] Rios I, Golab L, Keshav S. Analyzing the usage patterns of electric bicycles. In: Proceedings of the Workshop on Electric Vehicle Systems, Data, and Applications, EV-SYS 2016. Waterloo, Canada; 2016

[12] Hunecke M, Blöbaum A, Matthies E, Höger R. Responsibility and environment—ecological norm orientation and external factors in the domain of travel mode choice behaviour. Environment and Behavior. 2001;**33**:845-867

[13] Simons M, Van Es E, Hendriksen I. Electrically assisted cycling: A new mode for meeting physical activity guidelines? Medicine and Science in Sports and Exercise. 2009;**41**(11):2097-2102

[14] Directive 2002/24/EC of the European Parliament and of the Council of 18 March 2002 relating to the type-approval of two or three-wheel motor vehicles

[15] Kheirandish A, Kazemi M-S, Dahari M. Dynamic performance assessment of the efficiency of fuel cell-powered bicycle: An experimental approach. International Journal of Hydrogen Energy. 2014;**39**:13276-13284

[16] Cardinali L, Santomassimo S, Stefanoni M. Design and realization of a 300 W fuel cell generator on an electric bicycle. Journal of Power Sources. 2002;**106**(1)

[17] Mellino S, Petrillo A, Cigolotti V, Autorino C, Jannelli E, Ulgiati S. A life cycle assessment of lithium battery and hydrogen-FC powered electric bicycles: Searching for cleaner solutions to urban mobility. International Journal of Hydrogen Energy. 2016;**42**(3):1830-1840

[18] ISO 14040. International Standard. Environmental Management-Life Cycle Assessment-Principles and Framework. Geneva, Switzerland: International Organization for Standardization; 2006. Available from: www.iso.org

[19] International Reference Life Cycle Data System (ILCD) Handbook—General Guide for Life Cycle Assessment—Provisions and Action Steps. Luxembourg, Belgium: Publications Office of the European Union; 2010. ISBN: 978-92-79-15855-1

[20] Sonnemann G, Margni M. Life Cycle Management. Netherlands: Springer; 2015

[21] De Schryver AM, Brakkee KW, Goedkoop MJ, Huijbregts MAJ. Characterization factors for global warming in life cycle assessment based on damages to humans and ecosystems. Environmental Science and Technology. 2009;**43**:1689-1695

Chapter 2

Biogas Power Energy Production from a Life Cycle Thinking

Enrique Alberto Huerta-Reynoso,
Hector Alfredo López-Aguilar, Jorge Alberto Gómez,
María Guadalupe Gómez-Méndez
and Antonino Pérez-Hernández

Abstract

The purpose of this chapter is to present a generalized model for the construction of inventories for the production of electricity through biogas. This general framework can be adjusted to any power plant that uses biogas, since it complies with the main material and energy balances. This chapter describes the main technologies used in biogas power energy production, separating them into five main subsystems that integrate the general life cycle inventory, as well as the inputs and outputs considered in the development of the inventories. The life cycle assessment (LCA) of two types of plants is presented as study cases: (i) the biogas power energy generation with organic waste in landfills as substrate and (ii) the biogas power energy generation using dairy cattle manure as substrate. Both systems, in addition to using different types of substrate, present differences in their substages. It is concluded that the generation of studies of life cycle analysis of technologies facilitates decision makers, producers, and government agencies to develop and identify areas of opportunity from life cycle thinking.

Keywords: biogas, emissions, smart industry, sustainable energy, cleaner energies

1. Introduction

Energy is a critical factor for global decision-making. The energy supply is not only an important support in the daily anthropogenic activities, but also an important macroeconomic element [1]. Månsson [2] suggests a relationship between the energy sector of a country and the adaptation to changes, such as environmental degradation and the food price increase. This adaptation is carried out in a political, economical, and social context. Consequently, the energy topic has increased its relevance in international relations and dependencies. According to Overland [3], energy is a sensitive factor in globalization.

The development and the recent social growth are highly dependent on nonrenewable energies. This dependency has several disadvantages. Nonrenewable energies (NRE) are directly associated with climate change [4]. Since the last century, the atmospheric concentrations of CO_2 have increased from ~290 to 400 ppm in 2015 [5]. This period is related to an exponential increase in the energy demand. Another disadvantage is the resources depletion. Oil, for example, demands gradually more investment for few products. According to the Mexican government reports, the investment in oil extraction and exploration has increased 140% between 2004 and 2012; meanwhile, 30% less of the daily production initially perceived was obtained [6]. In general, both climate change and the resources depletion are issues of great importance at present [7]. This panorama incentivizes the investment for economic, scientific, and technological development for alternative and renewable energies.

Nowadays, several potential renewable energy sources are known. Its usage depends on its accessibility and transformation capacity. The energy obtained from biological sources, such as wood, crop residues, municipal waste, or even organic industrial waste, is called biomass energy [8]. This is widely used for heating and cooking activities [9] and is one of the oldest renewable energies. Biomass can be harnessed in several ways; energy sources can be obtained as many products, such as hydrogen, ethanol, methanol, and methane for transformation into mechanical energy and electricity. The biomass use comes with disadvantages. Overexploitation of biomass could damage natural areas by promoting the creation of monocrops to meet the energy demand [10]. However, the energy obtained from biomass waste could be a useful renewable energy source.

A product derived from the biomass fermentation is the biogas. It is obtained by a bacterial degradation denominated anaerobic digestion (DA) [11]. It is commonly used to obtain two main products: biogas to produce energy and digestate used for agricultural soil treatment.

The production of energy through biogas is a key element for future global projections. In Mexico, the potential for power energy generation through biogas is between 652 and 912 MW [6]. The use of renewable resources needs to be developed at an accelerated growth rate to meet global energy demand [12]. Its implementation must be successful too, oriented toward sustainability [13]. An applicable methodology to evaluate by this approach is the life cycle assessment (LCA).

Currently, the LCA has been applied to various energy production systems. The adoption of technologies such as biogas is generally promoted by environmental issues [14], specifically for waste disposal. The LCA methodology is an important tool in the use and implementation of anaerobic digestion for the generation and use of both biogas and soil improvers. Harder et al. [15] conducted a study of wastewater sludge treatment and integrated a quantitative microbial risk analysis into the LCA results. Dressler et al. [16] and Van Stappen et al. [17] carried out LCA studies of biogas production, the first about maize in three different areas of Germany; the second is about the installation of a farm biogas plant. Based on a sequential approach, both studies conclude on the importance of carrying out regional inventories for their application in decision-making. In China, Xu et al. [18] studied the generation of biogas from food waste. They found that the electrical consumption and transport of raw materials comes with the highest potential impacts. Moreover, Hijazi et al. in Germany analyzed 15 different biogas systems and found from life cycle thinking that the type of raw material is a key factor in environmental impacts. Furthermore, the spatial distribution of the plant and the management of by-products presented the highest environmental impacts. On the other hand, Huttunen et al. [14] identified that the use of biogas and the final use of digestate are the most critical points in the production of biogas in Finland. The construction of local inventories is a necessity to improve the LCA studies quality.

LCA is a useful tool for sustainability assessment in biogas systems. There are several studies of LCA in biogas. A general framework for biogas production has been established by other authors in independent studies. However, the recent studies focus on particular stage improvements or new technologies implementation.

In order to facilitate the construction of biogas power generation inventories, a general framework is desirable. This chapter presents the life cycle analysis of the generation of electrical energy by biogas, dividing the biogas power generation in individual subsystems. Two scenarios were considered for the elaboration of the ACV study, the use of biogas from the dairy corral excreta and the one from the municipal sanitary landfill.

2. A biogas background

Biogas is produced by anaerobic bacteria that degrade organic matter in four general stages: hydrolysis, acidification, acetic acid production, and methane production. The gas phase product of anaerobic digestion is named biogas and its yield depends significantly on the substrate (raw material). Biogas is composed by a mixture of 50–75% methane, 25–50% carbon dioxide, and 2–8% other gases (nitrogen, oxygen, hydrogen sulfide, among others). The percentage of methane in the biogas mixture is the main component for its use as an energy source; this also depends on the substrate that is used. **Table 1** shows the potential production of methane with different types of substrate, as well as its yield.

Before converting biogas into electricity by motor generators, the biogas must be purified by a desulfurization and drying process [11].

The requirements of the biogas quality depend on its different applications. In general, the costs of biogas purification are associated with the technology used and the location of the biodigestion system [20]. The choice of the most appropriate technology for the removal processes will depend on the use of this energy, as well as the compounds present in the biogas.

The toxic effect of H_2S, which is a colorless, flammable gas, has been documented, at levels of 0–5 ppm in the air, it can be easily detected; at concentrations higher than 10 ppm, it can affect human health, and at 600 ppm, it can cause death [21]. This gas is in the top five of pollutant compounds by Environment Canada's National Pollutant Release Inventory [22]. The main problems in the use of biogas, due to high concentrations of this gas, are the corrosion that damages the engines and the production of sulfur oxides from their combustion, whose emissions are subject to international regulations [23]. Therefore, desulphurization of biogas and its purification are necessary to increase the possible applications of this energy [24]. The main removal technologies for this compound are presented in **Table 2**.

The design of an optimal digester depends mainly on the characteristics of the substrate, as well as the amount of dissolved, volatile solids, biodegradability,

Substrate type	C:N ratio	Methane yield (m^3 CH_4/kg VS)	Methane production (m^3 CH_4/m^3)
Pig manure (solid)	7	0.30	48.0
Cattle manure (solid)	13	0.2	32.0
Poultry droppings (solid)	7	0.30	48.0
Garden wastes	125	0.20–0.50	NR
Fruit wastes	35	0.25–0.50	NR
Whey (from industry)	NR	0.33	0.15

NR = Not reported.

Table 1.
Biomass characteristics for biogas production [19].

Method	Advantages	Disadvantages
Biological with O_2/air, (by filter/ scrubber, /digestor)	Low investment and operating costs: electricity and calorific demand. It does not require any chemical products or additional equipment. Easy operation and maintenance.	The H_2S concentration remains high. The O_2/N_2 excess complicates an additional cleaning An over The overload of air produces an explosive mixture
$FeCl_3/FeCl_2/FeSO_4$, (in digestors)	Low investment cost: storage tank and dosification pump. Low heat and electricity demand. Easy operation and maintenance. Compact technique. Air absence in biogas.	Low efficiency. High operation costs (Iron salt). Changes in pH and temperature are not beneficial for the digestion processes. Difficulty in dosing.
Bed of $Fe_2O_3/Fe(OH)_3$ Steel wool covered with rust Wood chips or impregnated balls	High removing efficiency >99% Low investment cost	Water sensibility High operating costs. Exothermic regeneration: Wood chips ignition risks The reaction surface is reduced for each cycle. The dust released could be toxic.
Absorption by water	Low costs when there is water availability (nonregenerative) CO_2 is removed too.	High operation costs: high pressure, cold temperature. Difficult technique It could be presented obstructions in the absorption column.
Chemical absorption NaOH, $FeCl_3$	Low electricity demand Lower volume, less pumping, smaller vessels (compared to water absorption) low CH_4 loss.	High investment and operating costs. Difficult technique. Nonregenerative.
Chemical absorption $Fe(OH)_3$,Fe-EDTA	High efficiency in removing: 95–100% Low operation costs Small volume required Regenerative Low CH_4 losses	Difficult technique Regeneration by oxygen CO_2/H_2CO_3 (using EDTA) causes precipitation Thiosulphates accumulation
Membrane	Removing >98% The CO_2 is also eliminated	High maintenance and operation costs Complex
Biological filter	High remotion rate > 97% Low operating costs	An additional H_2S treatment is required The O_2/N_2 excess dificult an aditional cleaning
Adsorption by activated coal.	High efficiency High purifying rate Low operation temperature Compact technique High load capacity	High operating and investment costs CH_4 losses H_2O and O_2 are necessary to remove H_2S H_2O could occupate the H_2S role Regeneration at 450°C

Table 2.
Types of biogas purification technologies [25].

density, buoyancy of the solids and particle size [26]. Bioreactors can be classified as dry and wet. Some common configurations are: (i) dry batch reactors, (ii) continuously stirred tank reactors, and (iii) dry continuous reactors.

Technological advances have focused on new configurations, variants and modifications of conventional biodigesters. The anaerobic membrane reactors (AnMBR) have attracted attention in the field of research. This technology consists in the use of a membrane for the separation of solids and liquids inside the reactor, facilitating the handling of the effluent [27]. The membranes can be of different materials, in general they can be polymeric, metallic and ceramic, each one with its particular advantages [28]. The configuration of the use of these membranes varies according to the design of the reactor and the particular needs, from internal membranes, to membranes that operate by vacuum [29].

The production of electricity through biogas has been a notable increase in recent years. It is known that only in the European Union, the energy generated by biogas increased to 20,467.20 GWh in the period 2010–2013 [30]. Likewise, in other countries it has been found feasible to generate electricity through biogas. Arshad [31] carried out an economic study of the feasibility of generating energy using biogas from poultry residues. In his study he lists AD technology with a high potential to reduce Pakistan's local power deficit. Likewise, there are studies related to the application of improvements in order to increase the efficiency and feasibility of studies. Markou [32] presents an economic study where he uses the heat energy generated by the production of energy with the incorporation of greenhouses in the biogas plant, and they conclude that this modification contributes favorably from the economic point of view. In general, energy demand, as well as the need to search for new technologies, has favored the research and development of DA technology.

3. Biogas in the life cycle inventory process

The process of generating biogas, as well as its consecutive stages for the production of electrical energy, consists of several stages. **Figure 1** mentions the components for the design of a biogas utilization plant. Each of the stages can have a different design depending on the needs and the type of substrate to be fermented.

The development of the life cycle assessment can be separated into five main stages: (i) the collection of the substrate and its storage, (ii) pretreatment of the

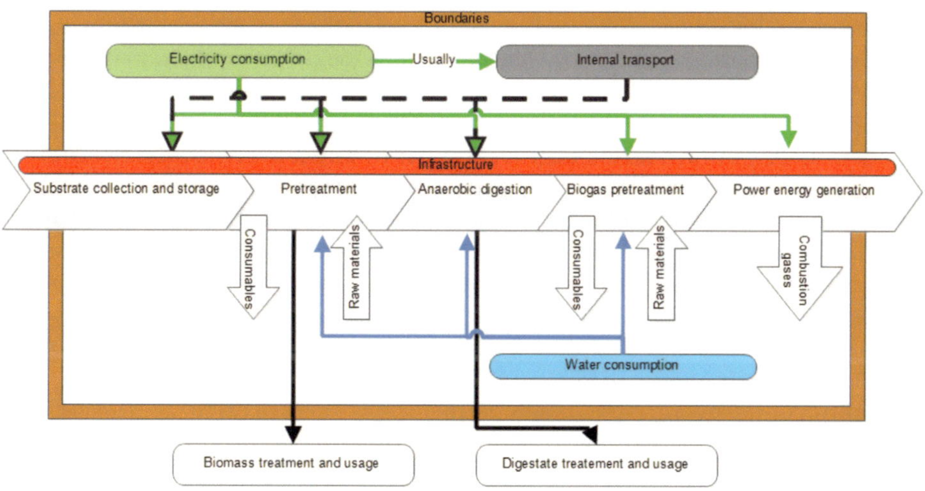

Figure 1.
Boundary limits for a biogas power plant life cycle inventory.

substrate, (iii) anaerobic digestion, (iv) pretreatment of biogas, and (v) generation of electric power. **Figure 1** shows a gate-to-gate flow diagram of the inputs and outputs of each of the stages of energy generation using biogas. The limits in **Figure 1** indicate the use of biomass and digestate in the pretreatment and anaerobic digestion stages, respectively, as well as outside the process of producing electricity through biogas. These processes can be considered by-products, which would allow them to be included within the limits as outputs to the technosphere. However, depending on the activities of the producers and the type of substrate, this digestate can be considered as waste. For a general analysis focused on the generation of electrical energy, these processes are considered as beside to the generation of energy, for the present work.

The essential part to be considered in the life cycle inventory (LCI) in a biogas plant is the infrastructure. Da Costa et al., mentioned infrastructure as part of the processes to consider in carbon dioxide emissions. Studies catalog the generation of emissions by infrastructure as low [33]. However, the contribution percentage of this system depends on the useful life of the plant because they are not constant emissions [32]. The maintenance, the configuration of the processes, and the type of bioreactor among other factors influence the useful life of the plant.

The first stage in the balance of matter and energy is the collection and storage of the substrate. It has been found that the variety of substrates used is wide. The types of substrate most used are: animal manure, agricultural residues, agroindustrial waste and even municipal organic waste [34]. These residues vary in composition, condition, density, as well as the type of collection and storage. In **Figure 1**, electricity was considered as an entrance; however, transportation plays a key role in this stage. Usually, substrates such as municipal waste are confined in landfills. The sealing of the cell favors anaerobic microbial consortiums that allow the generation of biogas. The same storage system fulfills the function of bioreactor, suppressing the pretreatment stage. For cases such as the use of animal waste, a more complex collection and storage system is necessary. These types of plants are usually of small or medium scale and are located near the source of the substrate. The collection of the substrate can be carried out by tractors or cargo vehicles. Also, transport and storage depend on the logistics, as well as the source of the substrate. The main factors to consider in this stage are the use of land and the emissions generated by transportation.

As mentioned in Section 2, the pretreatment stage varies according to the technology used. The access and availability of water are essential for the balance of material in this process. The relationship between the percentage of water and the content of solids in the substrate influences the yield and production of biogas [35]. Veluchamy and Kalamdhad [36] considered in their study a range of 80–85% humidity for an optimal methane yield in the DA of lignocellulosic substrate. Good practices mainly influence this stage and the use of water. Likewise, the electrical energy consumed is a key factor in this stage depending on the separation technology. Another important outlet is the residual organic matter. As already mentioned, this biomass can be considered as waste or as a byproduct depending on the use. This chapter focuses mainly on the generation of biogas and electricity. So the use of biomass and digestate (in the DA stage) were not considered.

The DA is the main stage of the biogas utilization plant. In addition to the infrastructure, it is necessary to consider the inputs and outputs in the monitoring and control of the parameters. Anaerobic digestion occurs in three main levels: psychrophilic (<25°C), mesophilic (25–45°C), and thermophilic (>75°C) conditions. Usually, digesters work at mesophilic conditions [34]. The energy consumption depends mainly on the temperature difference between the environment and the level of thermal conditions in which the plant works. Occasionally, producers opt for psychrophilic conditions due to the climatic conditions of the region [37]. Likewise, other parameters such as pH, micronutrients, and ammonia should be considered in the balance of inputs and outputs if necessary.

The pretreatment stage of the biogas is necessary for an optimal operation in the generation of electrical energy. As mentioned in Section 2, there are various techniques for removing unwanted components. The LCI depends on consumed inputs and the waste generated by the processes. So, the consumption of electricity and other energy inputs must also be considered in the balance.

The electric power generation stage is a key factor, not only for the construction of the LCI, but also for the design of the plant. The generation of electrical energy depends mainly on the technology used. It also depends on the composition of the biogas used. In this stage, the generation-consumption balance for the knowledge of net energy is crucial. The understanding of the energy flows consumed throughout the plant, compared with the energy generated, is critical for the optimization of the plant [38].

The five stages mentioned in **Figure 1** are the general scheme of an LCI for the generation of electric power. However, the configuration and stages may vary depending on the needs and the type of substrate. A system that uses urban solid waste sometimes lacks a pretreatment stage for the substrate. Also the collection methods may vary or belong to other linked operations. For example, the collection of animal excreta is a process also considered as cleaning stables on a farm. If the plant is in production, the collection is part of the cleaning system. The main outputs of the process are the residuals of the pretreatment stage of the substrate, the biogas, the digestate emitted by the anaerobic digestion and the emissions of combustion gases by the generation of electrical energy.

The development of the LCA is a comprehensive process. The form of construction of the inventory is explained in ISO 14044 [39]. **Figure 1** shows a gate-to-gate diagram of the system boundaries. However, the inclusion of other subprocesses should be considered according to the boundary conditions of the particular study.

4. Biogas and impact categories

In accordance with ISO 14040 [40], the life cycle impact assessment (LCIA) is intended to assess how significant the potential impacts of LCI emissions are. Generally, this evaluation is carried out through impact categories according to the emissions generated by the system. Currently, there are databases with impact categories already established. Some of the most used databases are ReCiPe, CML 2001, IMPACT 2002+, and IPCC 2013, in which the carbon footprint is obtained [41].

For a gate-to-gate study like the one shown in **Figure 2**, generally, the main emissions are air and water. However, depending on the system to be studied, additional emissions to the soil can be considered. For example, sometimes the digestate produced by the DA with a subsequent treatment can be used as a soil improver. However, if this is not carried out, it is possible to consider it as emission to the ground. This happens with the treatment and use of biomass in general. The consideration of these emissions within the limits of the system depends on the objective of the LCA.

The emissions to the air are carried out as a result of the generation of electrical energy by combustion, mainly. The exhaust gas mixture contains CO_2, CO, and H_2O, and other pollutant compounds such as SOx and NOx. The main impact of these emissions is the potential global warming. However, some of these compounds can generate from air toxicity to carcinogenicity.

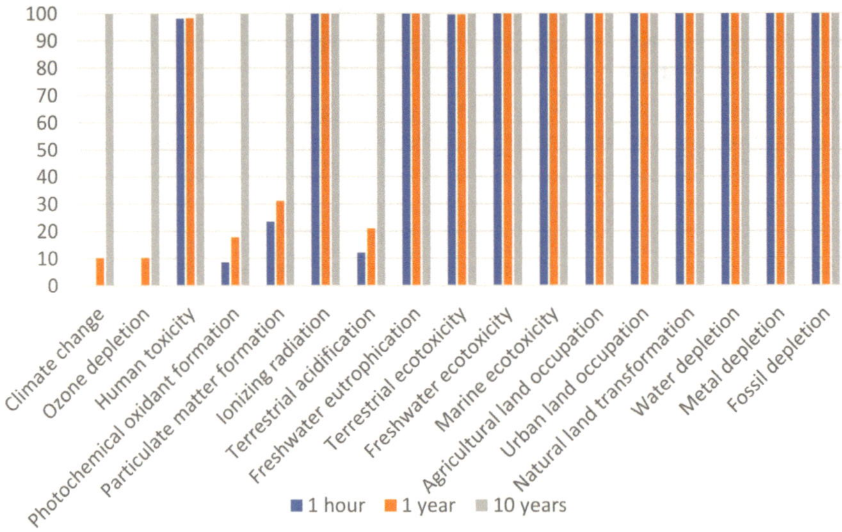

Figure 2.
Life cycle impact assessment of the landfill biogas power generation.

Regarding emissions to water, there is a considerable consumption of water, which is dependent on the technology and the size of the plant. As already mentioned, the composition of the digestate may contain micronutrients, which can be used for soil improvement. However, the emission into water of these compounds derived from phosphorus and nitrogen can cause eutrophication in water. Moreover, the emission of elements such as arsenic, lead, magnesium, etc. could contribute to the freshwater toxicity. It is advisable to carry out chemical analysis of the digestate at the exit of the system and in case of being disposed to bodies of water, consider it in the hydric balance.

The impact categories were selected based on the characterized emissions. For a standard biogas power plant, it is recommended to chose impact categories related to air toxicity and water contamination.

5. Study cases

In this section, two studies are presented. The first study case presents power generation from landfill organic matter biogas. The second study case presents the power energy generation from dairy manure biogas. Both studies are common examples of biogas-producing substrates.

5.1 Biogas from landfill

The electric power generation plant is located in the cd. Juárez, Chihuahua Landfill, road N °45, Juárez-Chihuahua (coordinates: 31° 33′15 "N 106 ° 29'33" W). The usage of different functional units allows to assess the sensibility of a system. For this study case, three separated periods of time were selected: (i) the generation of electricity from biogas during 1 hour of production, (ii) the annual production of electricity, and (iii) the generation of electricity for 10 years of production. Most of the inputs-outputs diagrams increase their scores linearly. However, it does not mean that the impact categories replicate this behavior.

The life cycle inventory is shown in **Table 3**. Inventories of the Ecoinvent v3.3 database were taken for the infrastructure of the plant [41]. It was considered a land use of 25,949 m^2. Furthermore, four generators with a generation capacity of 1230 kW of electricity were considered. These quantities were defined with the scale of the real power plant; however, the inventories were obtained from the Ecoinvent database. The construction of the plant was not considered due to the lack of information access. On the other hand, the air emissions provided by the producers were considered. The landfill biogas is generated by the anaerobic degradation of the organic matter. This process is carried out without any parameter control inside the landfill. For that reason, both biogas production and consumption is not considered for the study.

Products	**1 hour**	**1 year**	**10 years**	***Unit***
Functional unit	3296	2.89E+07	2.89E+08	kWh
Materials/fuels				
Infraestructura-BiogasJuarez/JRZ/MX	1	1	1	p
Emissions to air				
Carbon dioxide	2040	1.79E+07	1.79E+08	kg
Sulfur dioxide	2.79E-01	2.44E+03	2.44E+04	kg
Water	2.52E+02	2.21E+06	2.21E+07	kg
Ethane, 1,1,2,2-tetrachloro-	9.63E-05	8.43E-01	8.43E+00	kg
Ethane, 1,1-dichloro-	3.37E-04	2.95E+00	2.95E+01	kg
Ethene, 1,1-dichloro-	3.85E-04	3.37E+00	3.37E+01	kg
Ethane, 1,2-dichloro-	1.64E-02	1.43E+02	1.43E+03	kg
Benzene, 1,2,4-trimethyl-	1.01E-02	8.81E+01	8.81E+02	kg
Benzene, 1,2,3-trimethyl-	7.99E-03	7.00E+01	7.00E+02	kg
Benzene	2.91E-02	2.55E+02	2.55E+03	kg
Ethane, chloro-	9.63E-04	8.43E+00	8.43E+01	kg
Ethene, dichloro- (cis)	2.60E-03	2.28E+01	2.28E+02	kg
Methane, dichloro-, HCC-30	3.47E-03	3.04E+01	3.04E+02	kg
Benzene, ethyl-	5.94E-02	5.20E+02	5.20E+03	kg
m-Xylene	1.07E-01	9.38E+02	9.38E+03	kg
o-Xylene	4.94E-02	4.33E+02	4.33E+03	kg
Styrene	2.47E-02	2.17E+02	2.17E+03	kg
Ethene, tetrachloro-	1.61E-02	1.41E+02	1.41E+03	kg
Methane, tetrachloro-, CFC-10	7.70E-04	6.75E+00	6.75E+01	kg
Toluene	1.81E-01	1.59E+03	1.59E+04	kg
Ethane, 1,2-dichloro-	9.63E-05	8.43E-01	8.43E+00	kg
Ethene, trichloro-	5.68E-03	4.98E+01	4.98E+02	kg
alpha-Pinene	1.61E+00	1.41E+04	1.41E+05	kg
Limonene	6.58E+00	5.76E+04	5.76E+05	kg
P-cymene	8.13E-03	7.13E+01	7.13E+02	kg
Octamethyltetrasiloxane	1.29E-01	1.13E+03	1.13E+04	kg
Phenyltrichlorosilane	2.26E-02	2.26E-02	2.26E-02	kg

Table 3.
Life cycle inventory for each functional unit in the landfill biogas power plant.

The characterization for the life cycle impact assessment and the scenario comparison were calculated using SimaPro v8.5.2. **Table 4** and **Figure 2** show the results of the life cycle impact assessment of electric power generation in the Juarez biogas power plant. It can be seen in **Table 4**, a high score of emission equivalents in climate change and human toxicity categories. This is mainly due to the emissions of greenhouse gases and pollutants generated in combustion. On the other hand, emissions in both categories of human toxicity and marine eutrophication have indirect contribution like the equipment manufacture and the infrastructure.

Figure 2 shows the comparison of the potential impacts in the selected functional units. It can be seen that according to the increase of the time in the functional unit, the most sensitive categories scores are: climate change, ozone depletion, photochemical oxidation, particulate matter formation, and terrestrial acidification, which are mainly associated with air emissions. The results show a high sensitivity of gas emissions to the generation of electrical energy through biogas from landfill.

Moreover, the impact categories associated with soil such as terrestrial ecotoxicity, ionizing radiation, freshwater ecotoxicity, marine ecotoxicity, etc. (**Figure 2**) remain constant with increase in time in the functional unit. It is because of the secondary inventories, which are linked to the infrastructure and the manufacture of the power generators, whereby they are associated to indirect emissions.

According to **Figure 2**, air emissions are highly sensitive compared to other emissions. It is because of the biogas combustion caused by the power generation. Additionally, there are many volatile compounds generated with biogas produced in the landfill.

Impact category	Unit	1 hour	1 year	10 years
Climate change	kg CO_2 eq	4.53E+05	1.83E+07	1.79E+08
Ozone depletion	kg CFC-11 eq	1.32E-01	5.06E+00	4.94E+01
Human toxicity	kg 1,4-DB eq	3.07E+07	3.07E+07	3.13E+07
Photochemical oxidant formation	kg NMVOC	5.33E+03	1.09E+04	6.15E+04
Particulate matter formation	kg PM10 eq	1.50E+03	1.98E+03	6.38E+03
Ionizing radiation	kg U235 eq	8.87E+04	8.87E+04	8.87E+04
Terrestrial acidification	kg SO_2 eq	3.38E+03	5.82E+03	2.78E+04
Freshwater eutrophication	kg P eq	4.26E+02	4.26E+02	4.26E+02
Marine eutrophication	kg N eq	1.10E+03	1.10E+03	1.10E+03
Terrestrial ecotoxicity	kg 1,4-DB eq	1.37E+03	1.37E+03	1.37E+03
Freshwater ecotoxicity	kg 1,4-DB eq	9.64E+03	9.64E+03	9.64E+03
Agricultural land occupation	m^2a	4.68E+03	4.68E+03	4.68E+03
Urban land occupation	m^2a	3.13E+04	3.13E+04	3.13E+04
Natural land transformation	m^2	2.27E+02	2.27E+02	2.27E+02
Water depletion	m^3	9.38E+03	9.38E+03	9.38E+03
Metal depletion	kg Fe eq	2.89E+05	2.89E+05	2.89E+05
Fossil depletion	kg oil eq	2.89E+05	2.89E+05	2.89E+05

Table 4.
Life cycle impact assessment for each functional unit in the landfill biogas power plant.

5.2 Biogas from dairy manure

A well-known substrate for this activity is the waste of the livestock systems. Nowadays, there are several producers that use the manure of cattle for the generation of biogas [42–44].

The main objective of the LCA was to characterize the potential impacts of an electric power plant through biogas from dairy manure.

The study case scenario was a small-industrial generating plant with a generation capacity of ~ 22 kW, located in the dairy barn named "Establo Los Arados". The power plant is located in Meoqui, Chihuahua, Mexico (coordinates: 28° 14′35 "N 105 ° 28'14" W). The main activity is the dairy production; however, a biogas power plant was installed for both reduce operating cost in electricity consumption and managing the cattle manure generated.

In order to obtain the impact associated in 1 hour in the power plant, a functional unit of 22 kW h of electricity generation was selected. It is equivalent of the average power generated by the turbine installed. The boundary limits range from the manure collection to the power energy generation. These boundaries were defined based on the information access and the control parameters monitored in the power plant.

The harvesting system was divided into five main subsystems:

- Manure collection
- Barn infrastructure
- Biogas generation
- Power energy generation
- Power plant infrastructure

The pretreatment systems of both the substrate and biogas are included in the stage of biogas generation. Furthermore, a plant life of 30 years was assumed, considering a minor maintenance.

The unit processes and the boundary limits are illustrated in **Figure 3**. The water used in this process (blue line in **Figure 3**) is supplied by the barn. The green line (**Figure 3**) indicates the internal power supply. There are two different electricity sources considered: the municipal power supply and the biogas power energy generated.

The process of using biogas begins with the collection of manure. Because the power plant and the barn are in the same location, it is not necessary to travel long distances to transport the manure. The transport considers the route taken by the manure collector tractor, as well as the transport to the biogas production area. The continuous black line indicates the path of the substrate (manure). The substrate is transformed into the so-called stage of biogas production. This stage is separated into three substages: (i) pretreatment, (ii) anaerobic digestion, and (iii) purification. In this stage, waste is generated, such as solids (biosol) and effluent (biol). The biogas produced is taken to the stage of generation of electrical energy, which is incorporated into the supply line and for the self-consumption of the stable. In the stage of production of biogas, combustion gases are emitted, which were considered in the development of the inventories.

The operating conditions in the stages of the power generation, the water consumption and the energy consumption are information provided by the producers.

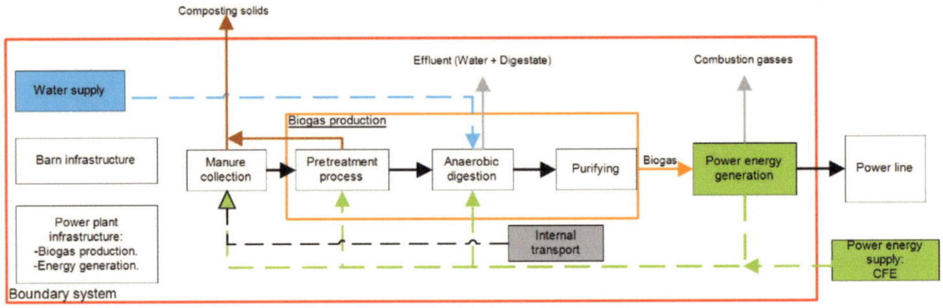

Figure 3.
Boundary limits for the dairy manure biogas power plant.

Literature information was included, mainly from the Ecoinvent database [41] and the parameters of the EPA-AP42 [45]. The literature information complemented the in situ measurements, which were carried out for both combustion gases (power energy generation) and effluent elements emitted (**Figure 3**).

The impact categories with a midpoint approach allow to assess the contribution of each of the systems with precision. **Figure 4** shows the percentage of contribution of each subsystem in the impact categories. This subsystem comparison allows to identify weak points and supports the technical decision making. Likewise, it allows to identify the direct impacts of production and the indirect impacts obtained from the consumption of resources.

Figure 4 shows a high contribution of infrastructure. It was identified that the electric power generation plant has an important effect in the categories related to depletion of resources, such as agricultural land occupation, ionizing radiation, urban land occupation, natural soil transformation, water depletion, metal depletion, and fossil depletion. It was found that the barn infrastructure contribute more than 80% of the total score in the categories of fossil depletion, freshwater ecotoxicity, and ozone

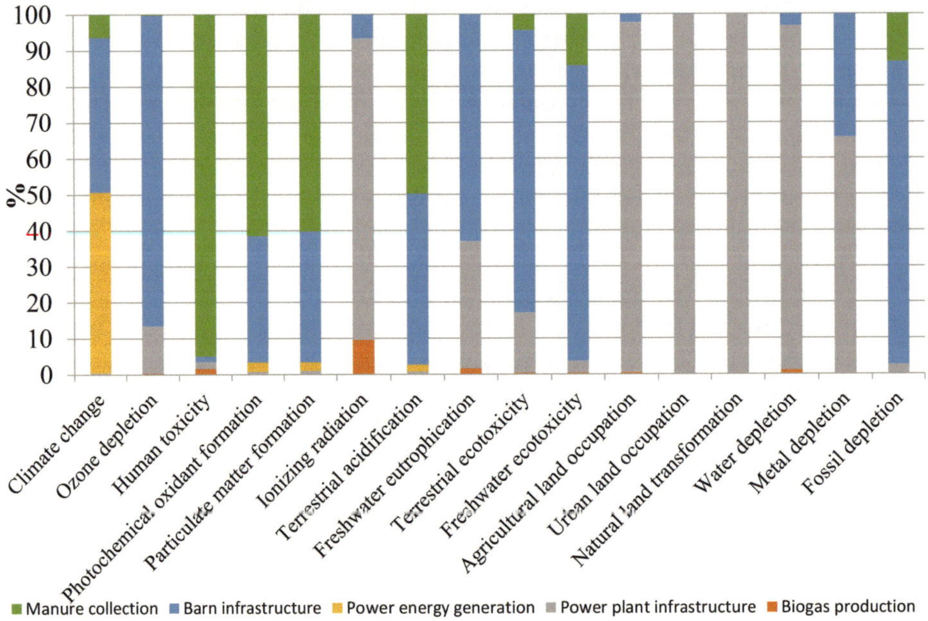

Figure 4.
Contribution for each impact category for biogas dairy manure power plant.

depletion. Likewise, it was considered maintenance of the barn was not relevant for the study. It was found that these categories show opportunity skills for decision-making.

The importance of the manure collection stage in the environmental load was identified. The usage of a machinery for transport and collection of excreta is difficult to modify due to the infrastructure adapted to the daily activities. However, it is possible to optimize the routes in the manure collection stage to mitigate the fuel consumption. With the appliance of this improvement, the environmental impact could be considerably reduced in the categories of human toxicity, photochemical oxidant formation, particle matter formation, and terrestrial acidification.

It was found that the generation of electrical energy is the main contributor to the climate change category due to the biogas combustion. Furthermore, the score in the categories of photochemical oxidant formation, particle matter formation, and terrestrial acidification is related to the generation of energy. However, this contributes to <10% of the total. So the power generation is the main opportunity skill in the category of climate change.

The identification of opportunity skills in the life cycle impact assessment allows stakeholders to make decision from a sustainable approach. On the other hand, in the life cycle assessment of the biogas landfill power generation, Section 5.1, sensitivity was analyzed by the comparison of different values in the functional unit to identify the most critical impact categories through the LCA.

6. Conclusions

Energy generation through biogas has gained relevance in recent years due to its potential capacity as a renewable energy source. An analysis of these technologies from the life cycle thinking is essential for sustainable development.

It was found that the separation of complex systems into subsystems or unit operations facilitates the development of inventories and the life cycle impact assessment. The infrastructure of the power plant initially implies an important contribution of potential impacts. However, with better practices and maintenance, better efficiency and useful life period, it mitigates the environmental impact.

The main impact categories (in the study cases) are related to the air emissions and water emissions. However, considering an efficient usage of the by-products, these emissions could be reduced. In the case of the power energy from biogas can be optimized if the by-products of the generation of biogas, like the digestate and solid phase inputs, are processed and conditioned to their usage as soil improvers. It reduces the environmental impact associated with the use of agrochemicals.

The LCA is a very useful tool for decision-making and environmental engineering. By using the general framework, any improvement in biogas power energy production could be incorporated in the system. In the study cases discussed in this chapter, the opportunity skills were detected, specifically, the combustion heat usage and the by-products coprocessing to mitigate the emissions. For future studies, more measurement data could be included.

Acknowledgements

This chapter was supported and funded by Centro de Investigación en Materiales Avanzados S. C., Universidad Autonoma de Chihuahua. The authors would like to thanks the Sectretaría de Energía in Mexico and Consejo Nacional de Ciencia y Tecnología (proyect SENER-CONACyT N°243715).

Conflict of interest

None declared.

Author details

Enrique Alberto Huerta-Reynoso[1], Hector Alfredo López-Aguilar[1], Jorge Alberto Gómez[2], María Guadalupe Gómez-Méndez[3] and Antonino Pérez-Hernández[1*]

1 Centro de Investigación en Materiales Avanzados, S. C., México

2 Universidad Autónoma de Ciudad Juárez, México

3 Universidad Autónoma de Chihuahua, México

*Address all correspondence to: antonino.perez@cimav.edu.mx

References

[1] Taylan O, Kaya D, Bakhsh AA, Demirbas A. Bioenergy life cycle assessment and management in energy generation. Energy Exploration & Exploitation [Internet]. 2017;**36**(1):014459871772587. Available from: http://journals.sagepub.com/doi/10.1177/0144598717725871

[2] Månsson A. Energy, conflict and war: Towards a conceptual framework. Energy Research & Social Science [Internet]. 2014 Dec [cited 2017 Jun 19];**4**:106-116. Available from: http://linkinghub.elsevier.com/retrieve/pii/S2214629614001170

[3] Overland I. Energy: The missing link in globalization. Energy Research & Social Science [Internet]. 2016 Apr [cited 2017 Jun 19];**14**:122-130. Available from: http://linkinghub.elsevier.com/retrieve/pii/S2214629616300093

[4] Bansal SK, Sreekrishnan TR, Singh R. Effect of heat pretreated consortia on fermentative biohydrogen production from vegetable waste. National Academy of Science Letters [Internet]. 2013;**36**(2):125-131. Available from: http://dx.doi.org/10.1007/s40009-013-0124-4

[5] Keenan TF, Prentice IC, Canadell JG, Williams CA, Wang H, Raupach M, et al. Recent pause in the growth rate of atmospheric CO_2 due to enhanced terrestrial carbon uptake. Nature Communications. 2016;**7**:13428

[6] Gobierno de la República. Reforma energética. México: 2013. pp. 1-24. Available from: http://reformas.gob.mx/wp-content/uploads/2014/04/Explicacion_ampliada_de_la_Reforma_Energetica1.pdf. [Accesed on August 2018]

[7] Hosseini SE, Wahid MA, Aghili N. The scenario of greenhouse gases reduction in Malaysia. Renewable and Sustainable Energy Reviews. 2013;**28**:400-409

[8] Cleveland CJ, Morris C. Dictionary of Energy (Expanded Edition)–Knovel [Internet]. Elsevier; 2009. Available from: https://app.knovel.com/web/toc.v/cid:kpDEEE0001/viewerType:toc/root_slug:dictionary-energy-expanded/url_slug:bioenergy?b-q=Bioenergy&b-subscription=TRUE&b-group-by=true&b-search-type=tech-reference&b-sort-on=default&issue_id=kt007STFA9

[9] Field CB, Campbell JE, Lobell DB. Biomass energy: The scale of the potential resource. Trends in Ecology & Evolution. 2008;**23**(2):65-72

[10] Peter C, Fiore A, Hagemann U, Nendel C, Xiloyannis C. Improving the accounting of field emissions in the carbon footprint of agricultural products: A comparison of default IPCC methods with readily available medium-effort modeling approaches. International Journal of Life Cycle Assessment [Internet]. 2016;**21**(6):1-15. Available from: http://dx.doi.org/10.1007/s11367-016-1056-2

[11] Gomez CDC. Biogas as an energy option: An overview. In: Wellinger A, Murphy J, Baxter D, editors. The Biogas Handbook: Science, Production and Applications. 2013. pp. 1-16

[12] Hijazi O, Munro S, Zerhusen B, Effenberger M. Review of life cycle assessment for biogas production in Europe. Renewable and Sustainable Energy Reviews. 2016;**54**:1291-1300

[13] Viktoria M. Examination of thermic treatment and biogas processes by LCA. Annals of Faculty Engineering Hunedoara–International Journal of Engineering. June, 2016;**7**(1):225-234. Available from: http://web.b.ebscohost.com/ehost/detail/

detail?vid=0&sid=d4cee639-182c-4bee-a5a1-2ad2fca30640%40pdc-v-sessmgr06&bdata=JnNpdGU9ZWhvc3QtbGl2ZQ%3d%3d#db=a9h&AN=96018304

[14] Huttunen S, Manninen K, Leskinen P. Combining biogas LCA reviews with stakeholder interviews to analyse life cycle impacts at a practical level. Journal of Cleaner Production [Internet]. 2014;**80**:5-16. Available from:http://dx.doi.org/10.1016/j.jclepro.2014.05.081

[15] Harder R, Peters GM, Molander S, Ashbolt NJ, Svanström M. Including pathogen risk in life cycle assessment: The effect of modelling choices in the context of sewage sludge management. International Journal of Life Cycle Assessment [Internet]. 2015;**21**(1): 60-69. Available from: http://dx.doi.org/10.1007/s11367-015-0996-2

[16] Dressler D, Loewen A, Nelles M. Life cycle assessment of the supply and use of bioenergy: Impact of regional factors on biogas production. International Journal of Life Cycle Assessment [Internet]. 2012;**17**(9):1104-1115. Available from: http://dx.doi.org/10.1007/s11367-012-0424-9

[17] Van Stappen F, Mathot M, Decruyenaere V, Loriers A, Delcour A, Planchon V, et al. Consequential environmental life cycle assessment of a farm-scale biogas plant. Journal of Environmental Management. 2016;**175**:20-32

[18] Xu C, Shi W, Hong J, Zhang F, Chen W. Life cycle assessment of food waste-based biogas generation. Renewable and Sustainable Energy Reviews [Internet]. 2015;**49**:169-177. Available from: http://dx.doi.org/10.1016/j.rser.2015.04.164

[19] Seadi T Al, Drosg B. Biomas resources for biogas production. In: The Biogas Handbook: Science, Production and Applications. First. Cambridge: Woodhead publishing; 2013. pp. 19-51

[20] Sun Q, Li H, Yan J, Liu L, Yu Z, Yu X. Selection of appropriate biogas upgrading technology–A review of biogas cleaning, upgrading and utilisation. Renewable and Sustainable Energy Reviews. 2015;**51**:521-532

[21] Rattanapan C, Boonsawang P, Kantachote D. Removal of H2S in down-flow GAC biofiltration using sulfide oxiding bacteria from concentrated latex wastewater. Bioresource Technology. 2009;**100**:125-130

[22] NPRI (National Pollutant Release Inventory). NPRI Pollutants in Canada 2002; 2002

[23] Dumont E. H_2S removal from biogas using bioreactors: A review. International Journal of Energy and Environment. 2015;**6**(5):479

[24] Kwaśny J, Balcerzak W. Sorbents used for biogas desulfurization in the adsorption process. Polish Journal of Environmental Studies. 2016;**25**(1):37-43

[25] Ryckebosch E, Drouillon M, Vervaeren H. Techniques for transformation of biogas to biomethane. Biomass and Bioenergy. 2011;**35**(5):1633-1645

[26] Murphy JD, Thansiriroj T. Fundamental science and engineering of the anaerobic digestion process for biogas production. In: Whellinger A, Murphy J, Baxter D, editors. The Biogas Handbook: Science, Production and Applications. Woodhead Publishing Limited. 2013. pp. 104-130. Available from: https://www.sciencedirect.com/science/article/pii/B9780857094988500051

[27] Liao B-Q, Kraemer JT, Bagley DM. Anaerobic membrane bioreactors: Applications and research directions. Critical Reviews in Environmental Science and Technology. 2006;**36**:489-530

[28] Lin H, Peng W, Zhang M, Chen J, Hong H, Zhang Y. A review on anaerobic membrane bioreactors: Applications, membrane fouling and future perspectives. Desalination [Internet]. 2013;**314**:169-188. Available from: http://linkinghub.elsevier.com/retrieve/pii/S0011916413000386

[29] Chen C, Guo W, Ngo HH, Lee D-J, Tung K-L, Jin P, et al. Challenges in biogas production from anaerobic membrane bioreactors. Renewable Energy [Internet]. 2016 Dec 1 [cited 2018 Sep 3];**98**:120-134. Available from: https://www.sciencedirect.com/science/article/pii/S0960148116302798

[30] Pablo-Romero M del P, Sïnchez-Braza A, Salvador-Ponce J, Sï¿½nchez-Labrador N. An overview of feed-in tariffs, premiums and tenders to promote electricity from biogas in the EU-28. Renewable and Sustainable Energy Reviews [Internet]. 2017;**73**(March 2016):1366-1379. Available from: http://dx.doi.org/10.1016/j.rser.2017.01.132

[31] Arshad M, Bano I, Khan N, Shahzad MI, Younus M, Abbas M, et al. Electricity generation from biogas of poultry waste: An assessment of potential and feasibility in Pakistan. Renewable and Sustainable Energy Reviews [Internet]. 2018;**81**(May 2017):1241-1246. Available from: https://doi.org/10.1016/j.rser.2017.09.007

[32] Markou G, Brulé M, Balafoutis A, Kornaros M, Georgakakis D, Papadakis G. Biogas production from energy crops in northern Greece: Economics of electricity generation associated with heat recovery in a greenhouse. Clean Technologies and Environmental Policy. 2017;**19**(4):1147-1167

[33] Mezzullo WG, McManus MC, Hammond GP. Life cycle assessment of a small-scale anaerobic digestion plant from cattle waste. Applied Energy. 2013;**102**:657-664

[34] da Silva OB, Carvalho LS, de Almeida GC, de Oliveira JD, Carmo TS, Parachin NS. Biogas–Turning waste into clean energy. In: Fermentation Processes [Internet]. 2017. pp. 161-180. Available from: https://www.intechopen.com/books/fermentation-processes

[35] Riya S, Suzuki K, Meng L, Zhou S, Terada A, Hosomi M. The influence of the total solid content on the stability of dry-thermophilic anaerobic digestion of rice straw and pig manure. Waste Management [Internet]. 2018;**76**:350-356. Available from: https://doi.org/10.1016/j.wasman.2018.02.033

[36] Veluchamy C, Kalamdhad AS. A mass diffusion model on the effect of moisture content for solid-state anaerobic digestion. Journal of Cleaner Production [Internet]. 2017;**162**:371-379. Available from: http://dx.doi.org/10.1016/j.jclepro.2017.06.099

[37] Massé DI, Saady NMC. Psychrophilic dry anaerobic digestion of dairy cow feces: Long-term operation. Waste Management. 2015;**36**:86-92

[38] Frank S, Nelles M. Energy flows in biogas plants: Analysis and implications for plant design. In: The Biogas Handbook: Science, Production and Applications. Cambridge: Woodhead Publishing; 2013. pp. 212-227

[39] ISO. ISO 14044:2006. Test 2006

[40] IMNC. NMX-SAA-14040-IMNC-2008. 2008

[41] Weidema BP, Bauer Ch, Hischier R, Mutel Ch Nemecek T, Reinhard J, Vadenbo CO, Wernet G. The Ecoinvent Database: Overview and Methodology, Data Quality Guideline for tthe Ecoinvent Database Version 3 [Internet]; 2013. Available from: www.ecoinvent.org

[42] Ledda C, Schievano A, Scaglia B, Rossoni M, Acién Fernández FG, Adani F. Integration of microalgae production with anaerobic digestion of dairy cattle manure: An overall mass and energy balance of the process. Journal of Cleaner Production. 2016;**112**: 103-112

[43] Hagenkamp-Korth F, Ohl S, Hartung E. Effects on the biogas and methane production ofcattle manure treated with urease inhibitor. Biomass and Bioenergy. 2015;**75**:75, 82

[44] Mathot M, Decruyenaere V, Stilmant D, Lambert R. Effect of cattle diet and manure storage conditions on carbon dioxide, methane and nitrous oxide emissions from tie-stall barns and stored solid manure. Agriculture, Ecosystems & Environment [Internet]. 2012;**148**:134-144. Available from: http://dx.doi.org/10.1016/j.agee.2011.11.012

[45] U. S. Environmental Protection Agency. 3.1 Stationary gas turbines. In: Office of Air Quality Planning and Standards, editor. Compilation of Air Pollutant Emission Factors [Internet]. 5th ed. U. S. Environmental Protection Agency; 1995. pp. 3.1-1-3.1-20. Available from: https://www3.epa.gov/ttn/chief/ap42/ch03/final/c03s01.pdf

Chapter 3

The Industrial Symbiosis of Wineries: An Analisys of the Wine Production Chain According to the Preliminary LCA Model

Agata Matarazzo, Fabio Copani, Matteo Leanza, Aldo Carpitano, Alessandro Lo Genco and Graziano Nicosia

Abstract

The circular economy refers to a term that defines an economy designed to be able to regenerate itself. Agri-food is one of the areas where the tools and strategies of the circular economy are implemented. The wine sector involving numerous stages of production and processing causes many impacts on the environment. Starting from the transport, to the distribution of wine products, there are several impacting processes on the environment. For the assessment of negative results although not of every product production process, the circular economy provides a more than suitable tool: the LCA (Life Cycle Assessment) has been implemented on the whole production chain of the product "Lenza di Munti", a bottle of wine by "Nicosia S.p.A.". The grapes used in the production of red wine are Nerello Mascalese and Nerello Cappuccio; instead of Carricante and Catarratto grapes used for white wine. This chapter provides a complete picture of the interactions between the product and the environment, to understand the environmental consequences and to provide the necessary information to define the best solutions.

Keywords: circular economy, LCA, agri-food, wine sector, Sicilian economy

1. Introduction

In recent years, along with the continuous development of the binomial environment industry, strong critical attention and assessment in the risks and impacts on the environment has matured. In the agri-food sector, it becomes very important to define carefully what are the environmental impacts and evaluate their weight, in order to improve or consolidate the quality of the production process. In the past, corporate objectives had little to do with the assessment of environmental impacts, rather they aimed almost exclusively at maximizing the quantity produced, remaining unrelated to everything that the production process caused. Today, it is essential to know more fully "what" is used in the production processes of agricultural products and "how" is used. In line with this trend, in the wine sector, in Italy, in recent years, we aim to produce quality wines, putting the quantity in the background; this allowed small producers to enter the market, thanks to the production of quality wines [1]. However, there is a change in trends regarding the size of producers: from small companies to large industrial companies. As a consequence this means more consumption, therefore higher emissions and impacts. The food industry disproportionately contributes to many global environmental problems. The production of wine is not an exception to the rule: it contributes to a variety of environmental burdens, mainly related to the use of pesticides and fertilizers in the vineyard and to the production of glass bottles. Target markets nowadays are more aware of ecological concerns than ever before; which means that costumers are increasingly more conscious of the products they are willing to buy and companies have to deal with this new context. The concept of sustainability is taken into account seriously in the Italian wine sector. The total vineyard area in Italy is approximately 656,000 hectares, and it is the third most important country per vineyard area in the world [2]. Moreover, the wine industry is an important contributor to the Italian economy, as it registers a sales volume of 9.5 billion euros. Sicily, in particular, widely contributes to the Italian wine industry: it is the region with the largest area of vineyards (for a total of 111,000 hectares), corresponding to around 17% of the overall Italian vineyard area and to 10.4% of the Italian total grapes production [3].

For this reason, this study will analyze a proactive company in the field, Nicosia S.p.A. To this end, the European Union has identified an instrument that evaluates and analyzes the entire life cycle of a product, from its cradle to its grave: the life cycle assessment. Thanks to this tool, companies can take steps toward the Green Economy road implemented by industrial symbiosis, also called industrial ecology. Industrial symbiosis, provided by industrial ecology, is an integrated management tool, which designs industrial infrastructures as if they were a series of interconnected ecosystems interfaced with the industrial ecosystem. It involves traditionally separate industries through an integrated approach aimed at promoting competitive advantages through the exchange of matter, energy, water, and/or by-products. The basis of this process is collaboration and the opportunity for synergy between companies. For example, in the case of Cantine Nicosia, the residual vinecce from the vinification process are reused for the production of spirits; this is the demonstration of how relationships, albeit simple, are able to constitute examples of industrial symbiosis. In particular, the study will focus on the inventory of a bottle of wine, the Lenza di Munti, by Nicosia S.p.A., in order to highlight any introduction in the process of any type of input, and to take note of the weight it has on the environment the output produced. The inventory is a real phase of the LCA, and in reference to the inherent standard [4, 5], this consists in the quantitative description of all the flows of materials and energy that cross the boundaries of the business system, both incoming and outgoing. Specifically, flow charts and data collection tables are used to prepare, with certainty of particulars, a complete and representative inventory of the product. In this study, each type of impact will be described, but above all that of CO_2 emissions. In this regard, many studies have highlighted what are the sources of CO_2 emissions in the wine sector, and they are: the phase of transport of inputs and outputs and waste management process and post process [6]. It will highlight the inter-company and inter-company functional relationships, representing those that are the subjects of the activity studied and the main actors, understanding those that are the bonds and relationships that exist between them.

The main actors involved in the activity studied are the company and its respective internal and external stakeholders including suppliers and customers.

The subject of the study is the wine sector and specifically the set of interactions that exist between the products and the environment and therefore the environmental impacts that arise from these interactions.

2. Materials and methods

As customers of goods and services we often don't think about all the consequences, it has got on what surrounds us: the environment and ecosystem in which we live are submitted to continuous stress and that determines its deterioration.

Companies of any dimensions are trying to modify their actions to handle their impact, to protect their reputation, and get prepared for stricter rules [5].

Using the LCA as an assessment tool of the impact of the wine sector sets some difficulties; in order to produce a bottle of wine several resources are needed, not only for the materials used for bottling but also we can have a great impact on the environment and also on mankind during the process of cultivation and treatment of grapes. Until a decade ago, producers used to think that a moderate use of pesticides and fertilizers gave a higher quality of wine. Indeed, during the last years, different choices have been made like leaving vines to their natural process without adopting solutions intrusive for the environment as the use of pesticides. In some particular cases, like Etna's vines, even irrigate is forbidden, if not in case of necessity, ensuring the natural growing process of grapes of wine.

The production sector begins by taking care of the vine which generally causes dangerous emissions neither for the environment nor for mankind unless pesticides are used; then during the harvest, it is time to harvest grapes in plastic containers with a useful life of 3 years. The harvest is then carried by trucks from the vine to the factory; once in there, the product is processed by a mechanic wine press which extracts juice and pulp from grapes creating in this way the must that has to be then transformed in wine. It is a particular and delicate operation that must be performed carefully. It is essential that the grapes have not been pressed by crushing during the transport because that could cause unwanted fermentation.

Subsequently or simultaneously with the pressing, the grapes are placed in a special automatic crusher-stemmer that separates the skeleton of the bunches, or the green part, from the berries. The result will be the destemmed crushed grapes, consisting of 80% of pulp, 15% of skins, and 5% of grape seeds, while the must has a composition of about 70–80% of water, 10–30% of sugars, and other substances.

Once the must is obtained, it is left to ferment in special containers; the aim of this phase is to transform the sugars into alcohol and give the wine the desired features and smells; depending on the type of vinification, in white or red, the skins are removed from the must (draining phase).

The duration of this phase varies from 5 to 15 days, and the main source of emission is that of CO_2 by the fermenter.

Once the fermentation days have passed, we proceed with the racking, in which the yeasts and the solid parts are separated from the wine. From this apparent waste, it is possible to obtain a second raw material, the marc, which is a primary part of the production process of some distillates such as grappa.

Once the racking is finished, the wine is subjected to the refining phase and at the right moment, the product is finished with the most invasive phase from an environmental point of view, which is bottling.

The use of LCA in environmental management and sustainability has grown in recent years as seen in the steadily increasing number of published papers on

LCA methodology and on case studies that have been performed to use LCA [7]. In many cases, these studies show that the most polluting phases of a food production system consist in the agricultural ones. Therefore, the research about innovations and environmental improvements should be addressed, above all, to the agricultural stages, also taking into consideration the economic feasibility [8]. Life cycle assessment is the factual analysis of a product's entire life cycle in terms of sustainability and allows the determination of input factors and output of the lifecycle of each product, evaluating the consequent environmental impact. This tool allows the phases where environmental issues are focused on to be identified as well as those who are responsible for the burden. LCA is a standardized methodology, which gives it its reliability and transparency. The standards provided by the International Organization for Standardization (ISO) describe the four main phases of an LCA: "goal and scope", "inventory analysis", "impact assessment," and "interpretation." These standards have been developed by the ISO/TC 207 Technical Committee and they are: UNI EN ISO 14040 (1998) [2, 3], by which the principles and the requirements for an LCA study are established; UNI EN ISO 14041 (1999), in which the objective and the field of application are defined and an inventory analysis is performed; UNI EN ISO 14042 (2000), that evaluates impacts: in particular, the main impact categories to be taken into consideration are the use of resources, human health, and ecological consequences; finally, with the UNI EN ISO 14043 (2000), improvements are analyzed. According to the ISO standards on LCA, it can assist in: identifying opportunities to improve the environmental features of products at various points in their life cycle; decision making in industry, governmental or nongovernmental organizations (e.g., strategic planning, priority setting, product and process design or redesign); selection of relevant indicators of environmental performance, including measurement techniques; and marketing (e.g., an environmental claim, eco-labeling scheme, or environmental product declarations). Lately, research about LCA shows how the concept of "economic sustainability" is spreading, and how helpful this tool is to achieve this objective. The agricultural sector provides a lot of examples about this usefulness: in Australia, for example, the GHG emissions from agricultural sector will continue to increase as Australia's agricultural export production is predicted to double over the next decade and the population is to reach 42.5 million by 2056. An LCA compiles the inputs and outputs from a production system, and in turn evaluates their potential environmental impacts, for example, GHG emissions [9]. Thus, the current research focuses on the determination of global warming impact or carbon footprint, that is, life cycle GHG emissions of horticultural products and identifies "hotspots" requiring mitigation strategies to reduce GHG emissions from the production and delivery of three important producers in Australia [10]. But it is not necessary to go far away from Catania; in less than 100 km, the environmental impacts of olive oil production have been studied by using the LCA. The Italian olive-growing sector has to face both the growing competition on the international olive oil market and the shift of the common agricultural policy (CAP) from market and price policies toward direct aids decoupled from production. A possible strategy to address this highly competitive scenario could be the renewal of olive groves through the adoption of innovative olive-growing models able to reduce production costs without worsening environmental sustainability. Two innovative olive-growing models are considered: the "High Density" (HDO) and the "Super High Density" (SHDO) olive orchards. From the environmental point of view, goal of the LCA is to build up the environmental profile of the two systems, in order to assess them and to identify their hot spots [8]. Depending on the complexity of the problem that a company wants to face, it is possible to choose between different types of LCA, each of which allows an in-depth solution to be found. These instruments are: LCA conceptual, preliminary LCA, and complete LCA. The first one is used only in the first phases, so that it often does not consider numerous aspects of the product lifecycle and does not make a comparison with other products. The second one, even if it does not consider the entire lifecycle of the product, allows the comparison between different products. The last one is the best methodology used to provide product improvements. A similar instrument to the LCA is the LCCA (lifecycle cost analysis). It is a methodology used for the economical evaluation of projects, in which the costs that come from owning, using, maintaining, and disposing of a certain product are considered vital to take a decision. In the end, this tool allows the determination of the overall cost of a certain product, considering its entire lifecycle [11].

3. Case study: Cantine Nicosia

Cantine Nicosia was founded in 1898 when Francesco Nicosia, the great-grandfather of the current owner, decided to open the first wine shop in Trecastagni, on the eastern side of Etna. The decisive entrepreneurial turnaround took place at the end of the twentieth century, thanks to the tenacity and innovative spirit of the current owner, Carmelo Nicosia, who, investing in the expansion and renovation of the vineyards and in the construction of a modern winery, will bring the family business to be the protagonist of the rebirth of Sicilian wine.

Today, Cantine Nicosia is a dynamic, modern, and efficient company, capable of looking to the future with full respect for tradition. The Trecastagni winery is the place where tradition combines with the most advanced technology. In the covered area of 4000 m^2, on a total surface of 27,000 m^2, between the wide winemaking area, the analysis laboratory, the modern bottling line, and the underground barrel cellar, the heart of the company pulsates.

Cantine Nicosia produces prestigious autochthonous wines, promoting the territory in full respect of the environment and enhancing the raw materials. In relation to this goal, the company has fully embraced the cause of organic and biological and sustainability, to combine environmental protection and food safety, with the aim of bringing to the tables of consumers a "zero impact" wine. Over the years, several milestones have been achieved in terms of international environmental certifications: from UNI EN ISO 9001:2015 to UNI EN ISO 14001:2015 Environmental Certification, from BRC Food & Beverage to the International Food Standard (IFS). Also recent is the achievement of an international "ethical" certification issued by SEDEX (Supplier Ethical Data Exchange), thanks to the environmental, biological, and vegan certifications. The company's future goal is the achievement of further certifications such as carbon footprint, water footprint, and LCA. Cantine Nicosia is a sustainable company actively involved in the protection of natural resources, and the optimal management of water resources is one of the essential points. Cellar work involves a high consumption of water, especially for cleaning and sanitation. In order to limit consumption, a system is implemented which, through the collection of both rainwater and reverse osmosis waste water, allows irrigation of farm lawns.

3.1 Functional unit

The subject of this study is a bottle of Lenza di Munti. It is necessary to list what the components of the bottle itself are: the glass bottle, the cork stopper, the label, and the aluminum cap. The supplier of glass bottles is O-I SALES. The cargo route starts from Vufflens-la-Ville (Switzerland), moves on to Aprilia (LT), and from there to Marsala (TP). At this point, the load will reach the Nicosia estates by road transport. The cork stoppers come from the Colombin company, located in Trieste. The company is certified: the used cork comes from Portuguese forests certified FSC. The transport of the caps takes place through the sea. The label is supplied by the company Moduli Continui s.r.l. (Padua), and transport is by road vehicles. The capsule is supplied by Enoplastic S.p.A. located in Bodio (VA); the delivery is carried out by road transport. Below, we will describe the uses and consumption of the listed parts. The final packaging consists of two elements: the cardboard containing the bottles and the packaging cellophane. The first is provided by the AICO company, Mascalucia (CT); the second, purchased in reels, is supplied by the company VIRAPACK, Acireale (CT). The Lenza di Munti is a wine intended for large retailers. Its winemaking process takes place in red and white. The first one is obtained from Nerello Mascalese and Nerello Cappuccio vines (80–20%); the second one is obtained from Carricante and Catarratto vines (60–40%). The analysis will focus on both vinification methods, generalizing the related consumption.

4. Results and discussion

Before moving on to the data collection, creating a flowchart in order to summarize and outline the production plant, in order to better understand which phases of the process we should analyze, according to our interests turned to be a useful practice. This study has, therefore, highlighted the main phases of the production plant, namely the ones which need an attentive analysis of the consumption (**Figure 1**; **Tables 1** and 2).

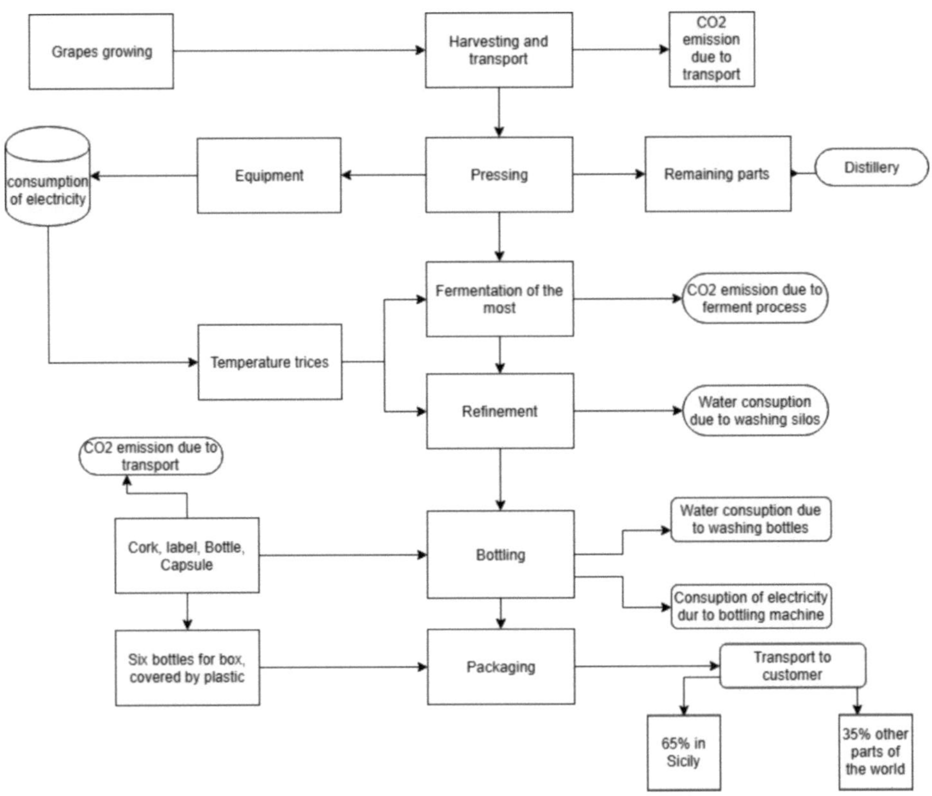

Figure 1.
Flowchart of the production plant.

Week activities	Per day	Per week	Per year
Microfiltration plant		3300.00	
Cross-flow filtration equipment (in Italian Filtro Tangenziale)		2700.00	
Total amount		12,000.00	624,000
Water consumption for winemaking (about 60 days per year)	4500.00		270,000.00
Total amount of the water consumption per year			2,376,000.000

Table 1.
Water consumption.

The data collection concerning any kind of consumption of the production plant makes it possible to make a list of the values about the main Nicosia S.p.A. water and energy consumption, noticing the related impacts. As far as water consumption is concerned, it is possible to distinguish consumption in three macro consumption areas:

Microfiltration plant;
Cross-flow filtration equipment;
General consumption of winemaking.

It is possible to notice that water consumption is closely connected to winemaking, since the daily consumption represents a larger amount than the week consumption of the other two activities which take advantage of water, namely microfiltration and cross-flow filtration equipment. Microfiltration is a filtration process, which removes solid particles from a fluid or from a gas, making them pass through a micro-porous membrane. Typically, the diameter of the small holes of these membranes is between 0.1 and 10 μm. In the field of winemaking, it is a process which is used in order to separate wine from all undesired substances; while, cross-flow filtration, which represents one of the most innovative processes in the field of filtration, improves the performance, not possible by any of the other processes concerning solid-liquid separation. In the field of drinks and winemaking,

Months	kW
Jan	33,680
Feb	34,499
Mar	36,453
Apr	43,353
May	37,309
June	41,259
July	36,792
Aug	30,139
Sep	29,307
Oct	30,457
Nov	29,241
Dec	33,141
Tot	415,630

Table 2.
Energy consumption.

this process aims at making perfectly transparent the product, stabilizing it microbiologically and, in case, it is necessary, eliminating partially colloides and oxidant enzymes, all in one phase. Doing so, all the following filtrations (included the centrifuge), pasteurization, and an improvement in the processes of stabilization of the product would be eliminated. This would lead to a lesser loss and manipulation of the product, saving on coadjuvants and on the cost of labour, with an overall reduction of the costs and a greater protection of the initial qualities of the product.

Concerning energy consumption, and, consequently, the related impacts, this study as made a list of the kW consumed per month by the production plant. The consumption values highlighted are between 29,000 and 43,000 kW (specifically, the month with the minor consumption was November, while the one with the most consumption was April). The total amount of energy consumption is high and amount to about 413,000 kW per year.

In order to finish the product, thus producing the bottle of wine ready to be delivered to the consumer, it is necessary to put together some components, coming from different geographical areas.

Consequently, this study cannot end just with the water and energy consumption. In order to have an overall view of how much impact the production of a bottle of wine has, this study has to calculate the kilometers which exist between the different components of the containers and the place where the product is bottled.

Inside an enterprise, the production of an item is not all produced by the enterprise; taking into consideration the case of Nicosia S.p.A., the main suppliers come from different regions and, in some cases, the first phase of the production of the item starts abroad (**Table 3**).

As far as the cork concerns, the raw material comes from Portugal, from certified FSC forests, that is an international certification, an independent and of third party, specific for forests and the products—wooden and nonwooden ones—derived from the forests. Specifically, not only is the forest certified but also the producer located in Trieste is, in particular for the chain of custody, the certification which ensures and makes the product trackable.

The transport of the bottle's components is made by tir from many parts of Italy to Trecastagni (CT). It is important to show the table of EURO Regulation referred to the emission limits of gas and fine dust. This case study takes into consideration an EURO 5 tir (**Table 4**).

The index (g/kWh) is a unit of measurement which could be considered as both an emission or a consumption index. In particular, it measures the grams of fuel for kW in an hour of the working system in full power.

In the years, the evolution of the EURO Regulation has entailed an increase of emission limits; in particular from EURO 5 to EURO 6, the limits of NOx emission

Component	From	km covered by the tir (km)	Hours of transport
Cork	Trieste	1437	15
Label	Padova	1256	13
Capsule	Bodio Lomnago Varese	1395	14
Bottle of glass	Marsala	352	4
Corrugated paper	Mascalucia	8.3	14 min
Plastics	Acireale	12.4	25 min
		4460.7	46 h and 39 min

Table 3.
Distance and fuel consumption.

EURO	NOx (g/kWh)	Particulate (g/kWh)
Euro 5	2.0	0.02
Euro 6	0.4	0.01

Table 4.
EURO norm.

Average of kW developed by a TIR	Total hours of transport	Emission of NOx	Emission of particulate and fine dust
330	46 h and 39 min	30,789	307.89

Table 5.
Transport emission.

passed from 2.0 to 0.4 g/kWh, and particulate emission in EURO 6 foresees 50% less than EURO 5 (**Table 5**).

It is easy to see that the major impact on the environment is due to the NOx and particulate emissions which increase the greenhouse effect.

5. Results and discussion

The quality of the waters is constantly monitored with analysis and controls. It also has a purifier for the waste waters of the cellar to re-insert clean water into the strata. The company also embraces three pivotal points on which the concept of sustainable development is based: economic, environmental, and social. Regarding the first aspect, the main aim is to increase the local economy through local suppliers and workers—a motivated choice also to reduce CO_2 emissions in a transversal way. The second aspect is the protection of the environment, which is why the terraces in the Monte Gorna vineyards as well as dampen the impact of the water—thus reducing the possible hydrogeological risks—enhance and protect the old technique of dry lava stone walls.

The choice of using cork stoppers is important. Thanks to this use, the demand is kept high, as following a lower demand for plugs, due to the greater use of alternative closures (synthetic and aluminum screw), there would be a sudden deforestation because the economic value of those trees it is only linked to the cork industry and to the production of corks. The cork, unlike alternative closures, is 100% sustainable, recyclable, and renewable, and it should not be forgotten that in the process of removing the bark, from which the corks are obtained, no tree is cut. A cork oak can live up to 250 years with decortic interventions every 9 years. Deforestation would also cause extensive damage to its huge ecosystem of 135 species of plants, 24 of reptiles and amphibians, 160 of birds, and 37 of mammals.

Last but not least is the environmental aspect. For 8 years, now the company has managed the care of the green in two roundabouts of the urban area of Trecastagni. The goal is to sensitize the population in the care of the common good.

6. Conclusion

It is estimated that winemaking industry involves an increase of gases in atmosphere:

152 kg of CO_2/ton (for the winemaking activity).

235 kg of CO_2/ton (for cellar activity).

Europe represents 60% of wine production in the world, in particular in Spain, Portugal, Italy, and France, which have a perfect climate allowing the growth of the grapes. The data of Nicosia S.p.A. represent an evident example of how it is possible to reduce impacts made by the wine production field over the environment. The company is proactive in all the chain of production. The solution is using innovative systems like cross-flow filtration equipment and cork certificated by FSC (Forest Stewardship Council).

It is clear too that there are some sources of air pollution, in particular the NOx and the particulate emitted by tir during the transports.

A solution could be using means of transport "eco-friendly" which pollute less, like tir powered by GPL, metan or electricity.

Author details

Agata Matarazzo[1]*, Fabio Copani[1], Matteo Leanza[1], Aldo Carpitano[2], Alessandro Lo Genco [1] and Graziano Nicosia[2]

1 Department of Economics and Business, University of Catania, Catania, Italy

2 Nicosia S.p.A., Catania, Italy

*Address all correspondence to: amatara@unict.it

References

[1] Iannone R, Miranda S, Riemma S, De Marco I. Improving environmental performances in wine production by a life cycle assessment analysis. Journal of Cleaner Production. 2016;**111**:172-180

[2] Roy P, Nei D, Orikasa T, Xu Q, Okadome H, Nakamura N, et al. A review of life cycle assessment (LCA) on some food products. Journal of Food Engineering. 2009;**90**(1):1e10

[3] Schimmenti E, Migliore G, Di Franco C, Borsellino V. Is there sustainable entrepreneurship in the wine industry? Exploring Sicilian wineries participating in the SOStain program. UNI EN ISO 9001:2015. Quality Management Systems. 2016

[4] ISO. ISO 14044—Environmental Management–Life Cycle Assessment–Requirements and Guidelines. Ginevra, Switzerland: International Organization for Standardization; 2006

[5] ISO. ISO 14044—Environmental Management—Life Cycle Assessment—Principles and Framework. In: International Organization for Standardization. Switzerland: Ginevra; 2006

[6] Amienyo D, Camilleri C, Azapagic A. Environmental impacts of consumption of Australian red wine in the UK. Journal of Cleaner Production. 2014;**72**:110-119

[7] Notarnicola B, Hayashi K, Curran M, Huisingh D. Progress in working towards a more sustainable Agri-food industry. Journal of Cleaner Production. 2012;**28**:1-8

[8] De Gennaro B, Notarnicola B, Rosella L, Tassielli G. Innovative olive-growing models: An environmental and economic assessment. Journal of Cleaner Production. 2011;**28**:70-80

[9] Greadel T, Allenby B. An introduction to life cycle assessment. In: Industrial Ecology. 2nd ed. Upper Saddle River, New Jersey, USA: Pierce Education; 2003

[10] Gunady MGA, Biswas W, Solah VA, James AP. Evaluating the global warming potential of the fresh produce supply chain for strawberries, romaine/cos lettuces (*Lactuca sativa*), and button mushrooms (*Agaricus bisporus*) in Western Australia using life cycle assessment (LCA). Journal of Cleaner Production. 2012;**28**:1-8

[11] Berners-Lee M, Howard DC, Moss J, Kaivanto KWA, Scott WA. Greenhouse gas footprint for small businesses-the use of input-output data. Science of the Total Environment. 2011;**409**:883-891

Chapter 4

End-of-Life Tire Destination from a Life Cycle Assessment Perspective

Thiago Santiago Gomes, Genecy Rezende Neto, Ana Claudia Nioac de Salles, Leila Lea Yuan Visconte and Elen Beatriz Acordi Vasques Pacheco

Abstract

Tires are complex materials manufactured from vulcanized rubber and various other reinforcing materials. One billion end-of-life tires (ELTs) are discarded annually, drawing attention from society. Options for their disposal include reuse, retreading, regeneration, co-processing, pyrolysis, and recycling; however, the ideal alternative has yet to be established. Life cycle assessment (LCA) has been used to quantify their impact and support the decision-making process, in order to determine the most beneficial alternative from an environmental standpoint. Scientific studies on LCA have been carried out on different continents, mainly Europe, Asia, and America. The aim of this chapter was to review studies on the life cycle assessment of end-of-life tire disposal. The main treatment and final destination options were reviewed as well as the most important limitations and aspects of the technologies studied. The most common form of disposal is recycling, with mechanical recycling for use in synthetic grass exhibiting the best environmental performance according to scientific research. Energy recovery also shows good performance, largely due to the emissions prevented through energy conversion. Co-processed and retreaded tires are regularly used for comparison but typically display poor environmental performance in relation to the first two alternatives.

Keywords: life cycle assessment, rubber, final destination, recycling, impacts, tire

1. Introduction

Approximately 1 billion unserviceable tires are discarded annually. The largest contributors are from the United States and the European Union, producing about 300 and 260 million, respectively [1–3]. Tires are a complex system containing 41% synthetic and natural rubber; up to 30 wt.% of additives such as silica and carbon black; 15 wt.% of reinforcing materials such as steel, polyester, and nylon; 6 wt.% of plasticizers and vulcanizing agents; and up to 2 wt.% of antiaging agents and other chemicals [4]. **Figure 1** shows the main components of a tire.

Selecting the final destination of tires requires significant knowledge and responsibility, since inappropriate disposal can result in a range of negative effects, including fires and the proliferation of mosquitoes. According to the waste hierarchy, there are several ways of disposing waste tires to mitigate environmental impacts, the most common being reuse, retreading, regeneration, co-processing, pyrolysis, and landfills [5, 6].

1.1 Reuse

Reuse involves using the whole tire or pieces of it to manufacture different rubber products for application in traffic and roadside barriers, the construction of parks and playgrounds, marine defense structures (dykes, wharfs, dams, and for coastal containment), channeling rainwater, artificial reefs, and biogas drainage [7, 8].

1.2 Reforming

Tire reforming can be achieved through three different processes, namely recapping, retreading, and remolding. All involve replacing one or more worn regions with crude rubber and submitting them to revulcanization to acquire the properties of a new tire. Recapping consists of replacing the tread, retreading replaces both the tread and its shoulder, and remolding, also known as bead-to-bead retreading, involves replacing the tread, shoulder, and entire sidewall surface [9, 10].

Reforming is an interesting strategy for used tire recovery, since it promotes savings in iron, rubber, and petroliferous resources and minimizes the problems associated with the disposal of used tires [11, 12]. Reforming is used primarily in the truck tire market, which can be retreaded three or four times [13, 14]. Retreading also provides energy savings because the energy required to manufacture a new tire is around 2.3 times greater than that needed for retreading [14, 15].

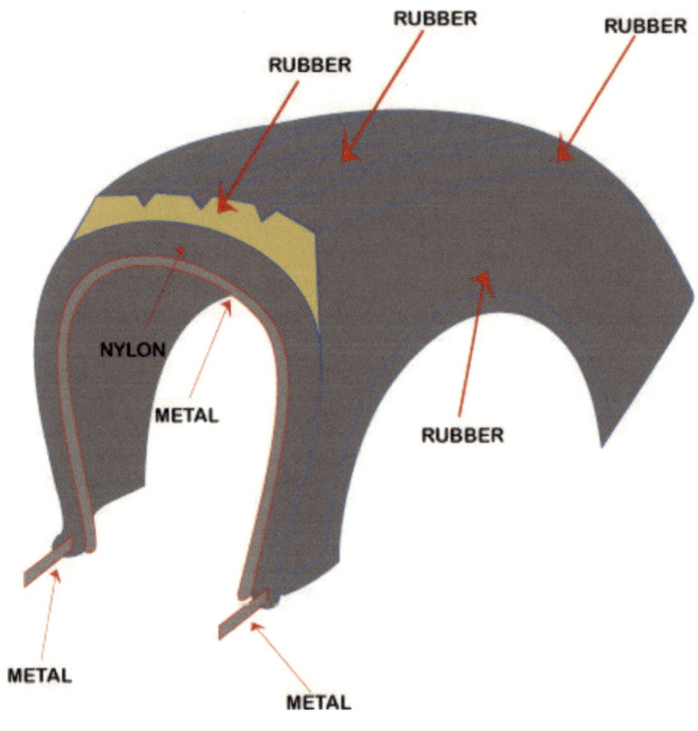

Figure 1.
Materials present in a tire.

1.3 Ground tire rubber

The presence of rubber and steel makes tire grinding a complex process. Rubber is an elastomeric material that requires special care, and steel has excellent mechanical properties, which hampers the molding process. Grinding can be carried out at ambient temperature, by ultrasound or cryogenically to produce small pieces of rubber for a variety of applications, including as a base for artificial grass pitches and playgrounds or an additive to asphalt [16, 17].

In grinding, vulcanized rubber is initially reduced to 7–10 cm particles that are placed into another grinder and processed at ambient temperature into smaller granules, removing steel (by magnetism) and fibers (using vibratory sieves and screens). Depending on the required product, additional processing (tertiary grinding) may be necessary to obtain even smaller particle sizes [17–19].

In grinding by ultrasound, whole tires are fed into a rotary grinder where ultrasound is generated, and the material is ground into 2 μm particles. The metal is removed by magnetic separators, and the final mixture consists of rubber and fabric [20, 21].

In cryogenic grinding, vulcanized rubber is first reduced to 50 mm particles in a mechanical pulverizer and then frozen at temperatures below −120°C in a cryogenic tunnel. The resulting rubber is fragile and can therefore be broken into small pieces in a mill. Metal and fibers are also removed, as occurs in mechanical grinding [18, 22–24].

1.4 Regeneration of tire rubber

In the case of regeneration, waste tires undergo chemical modification (degradation) in order to become more plastic, malleable, less viscous, and processable, that is, with properties similar to those of virgin rubber. Regeneration prompts the breaking of covalent carbon-carbon (C-C), carbon-sulfur (C-S), and sulfur-sulfur (S-S) bonds. If a number of C-C bonds are broken during the process, the main rubber chain may rupture, leading to serious structural disintegration [12].

The quality of products regenerated from waste tires varies according to their composition and the selectivity of the methods used in terms of the type and number of bonds to be broken. For regenerated waste to be deemed good quality, at least 70% of cross-linking must be carried out. It must also remain stable for at least 6 months and still be capable of being revulcanized at temperatures close to 170°C. Rubber regeneration can be carried out in the presence of a specific catalyst, which attacks the cross-linking points, or by applying enough energy to break these bonds. This process generally requires heat, chemical products, and mechanical energy. In principle, regeneration is used to obtain a product to replace virgin rubber with fewer technical requirements than the original product. Rubber is considered regenerated when it recovers its flow capacity and the characteristics of the original compound. Regenerated rubber can be used in carpets, furniture, asphalt mixtures, glues, and adhesives [25].

1.5 Co-processing in cement production kilns

Co-processing is defined as the use of waste materials to replace fuels and/or primary raw materials. Whole or ground tires are burned in a cement kiln to produce clinker, an intermediate product in cement manufacturing. The ash generated is not problematic because it is incorporated into the clinker, preventing the need for subsequent collection and treatment [16]. Silica and iron (contained in the tire) are used as secondary raw materials to replace sand and iron oxide in cement. The high temperatures (1500–1600°C) and oxidizing atmosphere in the cement production kiln allow complete combustion of the tire and almost total combustion of the volatile material produced during burning [7, 11].

The tires can be fed into the kiln whole or ground. Whole tires must be fed into the calcination zone of the kiln, while ground tires can be introduced into the burner zone [7, 20].

1.6 Co-processing in thermoelectric power stations

The use of fossil fuels (conventional power plants) in the form of coal, oil, and gas accounts for about 80% of the global energy demand [26, 27]. Nitrogen compounds and sulfur oxides produced by coal combustion have a significant effect on the environment and are responsible for acidification (acid formation) (HNO_3, H_2SO_4), increased ozone concentration at low altitudes, and high levels of particulate material [28, 29]. According to Singh et al. [30], using tires as a source material to generate energy in coal-fired power plants reduces NOx emissions and recovers the energy contained in the material. In this process, ground tires are combined with coal in the combustion unit to generate electrical energy. An important advantage of this process is that it lowers fossil fuel consumption [16]. Nevertheless, the energy conversion efficiency of power stations that use tires as raw material is 25–30% but far higher in conventional power stations. However, CO_2 emissions are around 23% lower when tires are used for energy generation [16].

1.7 Pyrolysis of tire rubber

Pyrolysis is a high-temperature chemical process that generates oil, gas, and carbon black. First, the tire is ground into 20 mm particles, fed into the pyrolytic reactor, and submitted to temperature (400–700°C) and pressure (0.01–0.04 MPa) conditions under which elastomers degrade. The products of the process consist of the following fractions: gaseous (hydrogen, methane, and carbonic oxides), liquid (water and oils), and residual solids (metals and dust) [16, 19].

An interesting process for the degradation of waste tires is thermolysis under pressure, which involves applying superheated steam and high pressure to obtain oligomers, gas, and liquid fuel. Used tires are placed into a preheating chamber (60–100°C), then fed into the reactor, and submitted to temperatures of 300–500°C and pressures of 1–1.2 atm. The resulting volatile hydrocarbons are removed and condensed, and the carbon residue is separated from the remaining metal [20, 31].

Another recycling technique for degrading tire rubber to obtain commercial products of interest is barodestruction, which is based on the pseudo-liquefaction of rubber at high pressure. Whole or ground tires are fed into the chamber at high pressure. The pseudo-liquefied rubber flows through the holes, and the nylon and metals are separated from the rubber. The metal is removed in the first step, and the rubber and nylon mixture is then passed through a grinder to separate the nylon. The gaseous emissions are treated using filters [20, 32].

1.8 Landfill disposal

This type of disposal consists of simply discarding tires in landfills, which is prohibited in Europe, according to Directive 2000/53/EU [33], and in countries such as Brazil [34]. In addition to shortening the useful life of the landfill, this practice impoverishes the soil, favors the proliferation of mosquitos, and makes the site prone to fires [7, 9, 15]. Fires caused by tires are difficult to extinguish. A tire

Authors and reference	Country	Impact method	Technology studied and/or process for end-of-life tires
Corti and Lombardi [18]	Italy	Ecopoint	• Combustion (waste-to-energy) • Substitution of fuel in cement clinker • Cryogenic pulverization • Mechanical pulverization
Ferrão, Ribeiro, Silva [15]	Portugal	Ecopoint	• Retreading • Recycling • Incineration (cement kiln) • Incineration (power plant) • Landfill
Li et al. [19]	China	Eco-indicator 99	• Ambient grinding • Devulcanization • Pyrolysis • Tire oil extraction
Clauzade et al. [7]	France	Not declared	• Recovery for retention basins • Tire recovery for infiltration basins • Mechanical recycling for steelworks • Mechanical Recycling in foundries application • Mechanical recycling for molded objects production • Mechanical recycling for synthetic turfs • Mechanical recycling for equestrian floors • Energy recovery for cement production • Energy recovery for urban heating
Fiksel et al. [16]	USA	Traci	• Cement production • Civil engineering • Incineration • Industrial boiler • Tire shredding and crumb production • Artificial turf • Molded products • Asphalt production • Retreading
Feraldi et al. [27]	USA	Traci	• Mechanical recycling • Energy recovery in co-incineration
Li et al. [35]	China	Eco-indicator 99	• Recycling to produce ground rubber
Sun et al. [36]	China	CML	• Recycling to produce reclaimed rubber
Ortíz-Rodriguez et al. [8]	Colombia	CML	• Reuse and retreading • Incineration • Grinding (recycling)

Table 1.
Summary of the studies assessed regarding LCA for waste tire rubber.

has around 75% of hollow space in relation to its entire volume, preventing these fires from being extinguished with water because the oxygen in this space feeds the fire. Additionally, the pyrolysis oil generated is a significant atmospheric, soil, and water pollutant [1, 2].

The most sustainable final destination for end-of-life tires is difficult to determine among the different possibilities available. The LCA tool has contributed to the decision-making process, requiring different technologies for each situation, region, and condition. As such, the aim of this chapter is to present studies that used LCA to investigate tire disposal options. Studies were reviewed by continent, and the environmental impact of each technology was evaluated.

The methodology used was divided into two stages. The first was to understand the different technologies applied for end-of-life tire disposal, and the second was to analyze life cycle studies that assessed these technologies in different parts of the world, including Europe, Asia, and America. To that end, a bibliographic review was conducted in different databases, such as ScienceDirect, Scopus, and Web of Science. The study selection criteria were directly related to the subject of the chapter, that is, end-of-life tire disposal based on life cycle assessment. The data from the selected articles are presented and summarized in **Table 1**.

2. Life cycle assessment of waste tire

2.1 General information on LCA

Life cycle assessment (LCA) can be used to quantify the impact of waste tire disposal and determine the most environmentally beneficial alternative for product manufacture and managing used products. LCA has also been applied to identify the most environmentally appropriate final destination for waste tires [3–7].

LCA can be applied to quantify the potential environmental impacts of a product and the resources used during its life cycle, including the acquisition of raw materials, production and use, and waste management. It can also be used to determine the best alternative for managing used products, encompassing their disposal, recycling, and reuse [37]. It is a broad assessment that considers all of the attributes or aspects of the natural environment, from human health to natural resources [38].

In order to standardize environmental management methodology, the International Organization for Standardization (ISO) developed the ISO 14.040 global standard [39], which defines the method for LCA application. An LCA study is divided into four phases: goal and scope definition, life cycle inventory (LCI) analysis, life cycle impact assessment (LCIA), and interpretation [40].

Defining the goal and scope includes establishing the motives for the study, the intended application, and target audience. The limits of the system under study are also described in this phase, in addition to defining the functional unit [40], which is a quantitative measure of the functions that the products (or services) perform. The results of the LCI provide information on the inputs (resources) and outputs (emissions) of the product during its life cycle in relation to the functional unit. The aim of the LCIA is to determine and evaluate the magnitude and significance of the potential environmental impacts of the system studied. In this stage, the functional units allow the relevant data to be compared. Inventory data are separated into midpoint [41] and endpoint (human health, ecosystem quality, and resource consumption) and converted into units via weighting factors for comparison [42]. Since the functional units have yet to be standardized, several names have been proposed, including Ecopoint unit. In this case, the values for each impact category are summed to produce a single value known as the Ecopoint, which corresponds to the environmental load of 1000 Europeans over a 1-year period [42, 43]. In the interpretation phase, the results of the previous stages are compared with the goal and scope in order to draw conclusions and provide recommendations [39].

In order to understand the state of the art, the papers developed in relation to the end-of-life tire destination that used the life cycle assessment were grouped by continents.

2.2 European LCA studies

Ferrão et al. [15] carried out an LCA of a new tire, whose life cycle phases were production, distribution, use, disposal, collection of the used tire, and recycling. The aim was to assess the impacts of a new tire during its life cycle as well as of four forms of recycling (recycling, retreading, fuel replacement, and energy generation) and disposal in a landfill. The Ecopoint approach was adopted, and the functional unit was a metric ton of used tires.

The results indicated that the most relevant phase in terms of environmental impacts was tire use. This was expected, since fossil fuels are the main fuel consumed during tire use and have a significant effect on the environment. Despite its impact, this phase is important in guaranteeing the safety of the vehicle, since the greater friction between the tire and the ground, the more secure the vehicle, but the more fuel it will consume [15].

Impacts resulting from landfill disposal are mainly related to the leaching of metals, stabilizers, flame retardants, and plasticizers, which are mixed with the rubber during tire manufacturing. Retreading is the most cost-efficient alternative in terms of the recovery of material and energy [11]. Although energy is consumed during retreading, consumption is 2.3 times greater when manufacturing a new tire. An important benefit of recycling is that it prevents the use of virgin material [15].

Burning whole tires to generate energy means they do not require grinding. However, a sophisticated burning system is needed to allow the use of high temperatures at specific points, and emissions must be kept within admissible limits [9]. Tire pyrolysis generates three products, namely, gas, oil, and carbon black. The energy potential of gas and oil (used to replace fuel) is similar to that of conventional products [44]. According to Van Beukering and Janssen [45], an important advantage of energy generation in cement kilns is that it does not produce solid residues and the sulfur emissions are not a significant problem because the sulfur generated is incorporated into the gypsum, which is added to the final product.

The results obtained in studies that applied LCA to analyze rubber recycling processes are detailed below. Corti and Lombardi [18] evaluated the following processes using LCA: mechanical pulverization, cryogenic pulverization, energy generation, and fuel replacement, the last applied in cement kilns. The functional unit was a metric ton of tires. The emissions generated were obtained via observations by the authors at different power plants, and average values were calculated. The only exception was the energy generation process, whose values were obtained from a thermodynamic model. The Ecopoint approach was adopted for the emission values.

Of the processes studied, cryogenic pulverization generated the most negative impacts due to its high water consumption when compared to the other processes. Other negative aspects include the greenhouse effect, eutrophication, and carcinogenic emissions (which were higher in cryogenic pulverization) [18].

The greenhouse effect, water consumption, and energy consumption were analyzed in greater detail. The impact on the greenhouse effect is assessed based on the equivalent CO_2 emissions into the atmosphere. According to Corti and Lombardi [18], cryogenic pulverization produces the poorest results, emitting around 450 kg of CO_2 equiv. per ton of tire processed. The energy generation process was the most beneficial because it consumes conventional materials, whereas fuel replacement showed no notable positive or negative influence on the greenhouse effect.

As previously mentioned, cryogenic pulverization displayed the most negative impact on water consumption, since it is high in the cryogenic step of this process. The energy recycling processes produced the best results, since water is not consumed to obtain energy.

In regard to energy consumption, cryogenic pulverization once again produced the worst results. As expected, much higher values were obtained in the two energy processes, given that they generate energy as opposed to consuming it.

Silvestraviciute and Karaliunaite [20] studied fuel replacement, mechanical pulverization, mechanical pulverization with ultrasound, thermolysis, and barodestruction recycling in terms of energy, atmospheric emissions, solid waste, and water consumption. The authors did not adopt a specific methodology for life cycle impact assessment, and only the values for each emission category were reported.

It can be concluded that the use of tires in the fuel replacement process is of significant interest in terms of energy; however, the emissions are similar to those produced using carbon as fuel [20]. In a study carried out by Corti and Lombardi [18], emission values were lower and negative, that is, the process did not result in new emissions. In the process studied by Corti and Lombardi [18], ground tires were added to the burner zone of the furnace, whereas Silvestraviciute and Karaliunaite [20] used whole tires added to the calcination zone. The advantage of the latter is the absence of the grinding step, since the whole tire is used; however, the drawback is that the metal is not recovered (during grinding, iron can be separated out and reused in another process).

In the process studied by Silvestraviciute and Karaliunaite [20], water consumption and solid waste generation were very low and not limiting factors. Gas and dust emissions are associated with fuel replacement and are zero or insignificant in the other processes.

Clauzade et al. [7] used LCA to assess used tire rubber as a substitute for different materials in a range of applications, including as a replacement for filler in retention dykes (concrete and polyethylene blocks) and infiltration (gravel substitute); as a filler at steelworks and foundries (to complement steel), in synthetic grass (instead of ethylene propylene diene copolymer–EPDM), at sports grounds (to replace sand), and in molded objects (instead of polyurethane); and as fuel for heating (coal substitute) and in cement plants (to replace fuel and raw materials). The study considered the transport of material from the generation center to the processing location, the impacts of the processes, and those prevented by the replacement. The authors concluded that reusing rubber as a filler for molded objects and synthetic grass provides the greatest environmental benefits. Additionally, the logistics of collection and transport is an important stage of the process.

2.3 Asian LCA studies

Li et al. [19] analyzed four processes for use in LCA: mechanical pulverization, regeneration, pyrolysis, and oil extraction. As in the studies mentioned above, the functional unit was 1 metric ton of tires. In accordance with Eco-indicator 99, disability-adjusted life years (DALY) were used to evaluate human health-associated impacts. The impact of one unit on this scale corresponds to the loss of 1 year of life. The unit used for ecosystem quality was the potentially disappeared fraction of species (PDF), in the form of PDF*m^{2}*yr (where m^2 is an area in square meters and yr, a year). An impact value of 1 for this unit indicates that all species within one square meter disappear over a year. For the resources category, the unit used was MJ of surplus energy, where an impact value of 1 indicates that an area previously used to extract resources requires 1 MJ of additional energy in order to be used again due to the decline in the natural resources available [46].

The following impacts were considered in the present study: ecotoxicity, acidification and nitrification, emission of carcinogenic materials, global warming potential, emissions of inorganic and organic materials harmful to human health, and the consumption of fossil fuels.

Global warming is caused primarily by the emission of CO_2, CO, N_2O, and CH_4. This study [19] found that only the oil extraction process caused negative effects. The processes that obtained the best environmental performance were mechanical pulverization and pyrolysis. The effects of the first three processes (mechanical pulverization, regeneration, and pyrolysis) are negligible when compared to oil extraction, which uses carbon as an energy source and generates large amounts of heavy metals.

The impacts assessed in the ecotoxicity category were those related to heavy metal and aromatic compound levels in the soil or air. Once again, oil extraction had the most negative impact because carbon is burned as an energy source.

In relation to fossil fuel consumption, all the processes obtained negative values because virgin material was not required, precluding the need for energy consumption during extraction. Even in oil extraction, fuel consumption is avoided, since the oil generated is an energy source [35].

The predominant management option in the Chinese end-of-life tire market is the production of ground rubber [35] for regeneration. In order to improve the environmental performance of ground rubber production, Li et al. [35] made a series of technical recommendations based on the Eco-indicator 99 method. The process consists of three main stages: ground rubber preparation, regeneration, and refining.

According to the authors [35], respiratory inorganics obtained the most severe results, that is, the highest relative contribution among the other impact categories assessed. With respect to regeneration, devulcanization was responsible for most of the environmental loads, corresponding to 66.2% of the total impact. Moreover, improvements in the flue gas treatment contributed to better performance. The use of renewable and clean energy can improve environmental performance by approximately 22%. These results could be used as a guide to reduce the environmental load when producing ground rubber from scrap tires. Moreover, increasing energy efficiency, improving environmental protection equipment, and using clean energy are effective measures to achieve this goal [35].

Still in regard to the Chinese tire market, Sun et al. [36] assessed the environmental impacts of radial tires for passenger vehicles. The authors used the CML method to analyze raw material extraction, tire production, use, and end of life. However, they considered only five out of eight impact categories, namely global warming potential (GWP), acidification potential (AP), photochemical oxidant creation potential (POCP), eutrophication potential (EP), and human toxicity potential (HTP), since these are easier to explain and based on direct emissions that are easy to correlate, in addition to being more important to tire production.

It was assumed that all end-of-life tires were collected and recycled and that, after separating the different tire components, the rubber was completely regenerated to replace synthetic rubber. This recovery and recycling process only showed negative impacts for GWP, EP, and HTP, meaning it prevents emissions as opposed to causing them. However, the main environmental impacts observed during the production of reclaimed rubber and waste treatment were for AP and POCP [36].

2.4 American LCA studies

Fiksel et al. [16] studied fuel replacement, energy generation, retreading, and mechanical grinding. The grinding process analyzed was aimed at the application of rubber in civil construction (as asphalt and a base for synthetic grass) and as a filler in new products. The authors found that using waste tires as raw material for synthetic grass is the most promising alternative, followed by energy recovery (co-processing in cement kilns and energy generation). However, the study was conducted in the United States, where the market for artificial grass is saturated, and, as such, they concluded that energy recovery is currently the most viable alternative.

Feraldi et al. [27] evaluated two final destinations for tires in the United States: grinding and energy recovery. The authors used the TRACI method and analyzed the future prospects for tire disposal considering changes in US energy matrix. The results identified grinding as the ideal final destination, given that energy recovery involves burning and emission of harmful compounds. With regard to future prospects, the authors concluded that the reduction in the impacts of each process would be negligible.

In Colombia, Ortíz-Rodríguez, Ocampo-Duque, and Duque-Salazar [8] used LCA to estimate the environmental impacts of three different alternatives for tires at the end of their useful lives in a case study at the Valle del Cauca Department. The first option was reuse and retreading, the second incineration, and the third grinding to obtain new products. CML-2001 was used to calculate the environmental impact indicators.

Grinding to manufacture flooring and rubber incineration in cement plants exhibited the best environmental results, largely because they prevent harmful effects by recovering the material. Comparison of the different waste tire recovery and disposal processes indicated that retreading and the production of multipart asphalt displayed the worst environmental performance. The performance categories used were global warming potential, ozone layer depletion, acidification, abiotic resource depletion, and photochemical ozone formation. The phases that most contributed to the recovery process were fuel consumption, initial synthetic rubber production, and conversion into liquid asphalt [8].

3. Summary of the studies evaluated

A comparison of the papers presented in **Table 1** shows that the studies are concentrated in Europe [7, 15, 18], the United States [16, 27], and China [19, 35, 36]. With respect to different forms of disposal, it is noteworthy that earlier studies describe a larger number of options, while current research focuses on comparing alternatives to recycling, as well as exploring different applications and recycling techniques [8, 35].

There is no consensus regarding the best impact method for tire recovery studies, although regional preferences are observed. European studies showed a preference for Ecopoint [15, 18], while American papers used only the TRACI method [16, 27], Chinese authors applied both Eco-indicator 99 and CML, and Colombian studies the CML [8, 19, 35, 36].

It is important to underscore that more LCA studies are needed to better understand the impacts of alternatives to traditional tire management, particularly when tires are submitted to new industrial processes, such as recycling [21, 47, 48].

4. Final considerations

End-of-life tire disposal was shown to be of great interest in Europe, Asia, and America, as a means of contributing to the decision-making process in selecting the best technological alternative from an environmental standpoint. Studies demonstrated that the best environmental performance, in general, was mechanical recycling for use in synthetic grass. The worst environmental performance was observed in co-processed and retreaded tires. There is no consensus regarding the best tire recovery method, although regional preferences are observed. European studies showed a preference for Ecopoint, while their American counterparts prefer Traci methodology for life cycle assessment.

Acknowledgements

The authors would like to thank the National Council for Scientific and Technological Development (CNPq) and the National Council for the Improvement of Higher Education (CAPES).

Author details

Thiago Santiago Gomes[1], Genecy Rezende Neto[1], Ana Claudia Nioac de Salles[2], Leila Lea Yuan Visconte[1,3] and Elen Beatriz Acordi Vasques Pacheco[1,3]*

1 Mano Institute of Macromolecules, Federal University of Rio de Janeiro, Rio de Janeiro, Brazil

2 Fraunhofer Institute for Chemical Technology, Pfinztal, Germany

3 Environmental Engineering Program, Federal University of Rio de Janeiro, Rio de Janeiro, Brazil

*Address all correspondence to: elen@ima.ufrj.br

References

[1] Rezende NG. Inventário do ciclo de vida comparativo de dois processos de regeneração de borracha de pneus inservíveis. Thesis. Federal University of Rio de Janeiro (UFRJ); 2013

[2] Thomopoulos N. Global Market for the Tire and Rubber Remediation and Recycling Industry: AVM155A|BCC Research [Internet]. bcc Research. 2018 [cited 2018 Aug 27]. p. 260. Available from: https://www.bccresearch.com/market-research/advanced-materials/global-market-for-the-tire-and-rubber-remediation-and-recycling-industry-avm155a.html

[3] Goldstein Research. Global Tire Recycling Market Share, Size, Trends Analysis | Forecast 2024 [Internet]. 2018 [cited 2018 Aug 27]. p. 275. Available from: https://www.goldsteinresearch.com/report/global-tire-recycling-industry-market-trends-analysis

[4] Continental Reifen Deutschland GmbH. Tyre Basics: Passenger Car Tyres. TDC 10/2013. [Internet]. 2013. p. 11. [cited 2018 Aug 27]. Available from: https://blobs.continental-tires.com/www8/servlet/blob/160650/d2e4d4663a7c79ca81011ab47715e911/download-tire-basics-data.pdf

[5] Van Ewijk S, Stegemann JA. Limitations of the waste hierarchy for achieving absolute reductions in material throughput. Journal of Cleaner Production. 2016;**132**:122-128

[6] Silva da ECR. Estudos das tecnologias de destinação de pneus inservíveis através de avaliação de impactos ambientais para um desenvolvimento sustentável. Thesis. Federal University of Rio de Janeiro (UFRJ); 2004

[7] Clauzade C, Osset P, Hugrel C, Chappert A, Durande M, Palluau M. Life cycle assessment of nine recovery methods for end-of-life tyres. International Journal of Life Cycle Assessment. 2010;**15**(9):883-892

[8] Ortíz-Rodríguez O, Ocampo-Duque W, Duque-Salazar L. Environmental impact of end-of-life tires: Life cycle assessment comparison of three scenarios from a Case Study in Valle Del Cauca, Colombia. Energies. 2017;**10**(12):2117

[9] Jang J-W, Yoo T-S, Oh J-H, Iwasaki I. Discarded tire recycling practices in the United States, Japan and Korea. Resources, Conservation and Recycling. 1998;**22**:1-14

[10] Simic V, Dabic-Ostojic S. Interval-parameter chance-constrained programming model for uncertainty-based decision making in tire retreading industry. Journal of Cleaner Production. 2017 Nov 20;**167**:1490-1498

[11] Amari T, Themelis NJ, Wernick IK. Resource recovery from used rubber tires. Resources Policy. 1999;**25**

[12] Cui X, Zhao S, Wang B. Microbial desulfurization for ground tire rubber by mixed consortium-Sphingomonas sp. and Gordonia sp. Polymer Degradation and Stability. 2016 Jun 1;**128**:165-171

[13] Beukering PJH, Janssen MA. Dynamic integrated analysis of truck tires in Western Europe. Journal of Industrial Ecology. 2000 Apr;**4**(2):93-115

[14] Cooper DR, Gutowski TG. The environmental impacts of reuse: A review. Journal of Industrial Ecology. 2017;**21**(1):38-56

[15] Ferrão P, Ribeiro P, Silva P. A management system for end-of-life tyres: A Portuguese case study. Waste Management. 2008;**28**(3):604-614

[16] Fiksel J, Bakshi BR, Baral A, Guerra E, Dequervain B. Comparative life cycle assessment of beneficial applications for scrap tires. Clean Technologies and Environmental Policy. 2011;**13**(1):19-35

[17] Myhre M, Saiwari S, Dierkes W, Noordermeer J. Rubber recycling: Chemistry, processing, and applications. Rubber Chemistry and Technology. 2012 Sep 8;**85**(3):408-449

[18] Corti A, Lombardi L. End life tyres: Alternative final disposal processes compared by LCA. Energy. 2004;**29**:2089-2108

[19] Li X, Xu H, Gao Y, Tao Y. Comparison of end-of-life tire treatment technologies: A Chinese case study. Waste Management. 2010;**30**(11):2235-2246

[20] Silvestravičiūtė I, Karaliūnaitė I. Comparison of end-of-life tyre treatment technologies: Life cycle inventory analysis. Environmental Research, Engineering and Management. 2006;**35**(1):52-60

[21] Dobrotă D, Dobrotă G. An innovative method in the regeneration of waste rubber and the sustainable development. Journal of Cleaner Production. 2018;**172**:3591-3599

[22] Lagarinhos CAF, Tenório JAS. Tecnologias utilizadas para a reutilização, rcciclagcm e valorização energética de pneus no Brasil technologies for reusing, recycling and energetic valorization of tires in Brazil. Polímeros Ciência e Tecnol. 2008;**18**(2):106-118

[23] Laurent A, Clavreul J, Bernstad A, Bakas I, Niero M, Gentil E, Christensen TH, Hauschild MZ. Review of LCA studies of solid waste management systems - Part II: Methodological guidance for a better practice. Waste Management. 2014;**34**(3):589-606

[24] Thomas BS, Gupta RC. A comprehensive review on the applications of waste tire rubber in cement concrete. Renewable and Sustainable Energy Reviews. 2016 Feb 1;**54**:1323-1333

[25] Pacheco EBAV, Visconte LL, Furtado CRG, Alburquerque Neto JR. Recycling of rubber: Mechano-chemical regeneration. In: Advances in Materials Science Research. New York: Nova Science Publishers; 2012

[26] Muneer T, Asif M, Munawwar S. Sustainable production of solar electricity with particular reference to the Indian economy. Renewable and Sustainable Energy Reviews. 2005;**9**:444-473

[27] Feraldi R, Cashman S, Huff M, Raahauge L. Comparative LCA of treatment options for US scrap tires: Material recycling and tire-derived fuel combustion. International Journal of Life Cycle Assessment. 2013;**18**(3):613-625

[28] Webb AH, Hunter GC. Power-station contributions to local concentrations of NO, at ground level. Environmental Pollution. 1998;**102**:283-288

[29] Liu Z, Guan D, Wei W, Davis SJ, Ciais P, Bai J, et al. Reduced carbon emission estimates from fossil fuel combustion and cement production in China. Nature. 20 Aug 2015;**524**(7565):335-338

[30] Singh S, Nimmo W, Gibbs BM, Williams PT. Waste tyre rubber as a secondary fuel for power plants. Fuel. 2009;**88**:2473-2480

[31] Zeaiter J, Azizi F, Lameh M, Milani D, Ismail HY, Abbas A. Waste tire pyrolysis using thermal solar energy: An integrated approach. Renewable Energy. 2018 Aug 1;**123**:44-51

[32] Choi G-G, Oh S-J, Kim J-S. Scrap tire pyrolysis using a new type two-stage pyrolyzer: Effects of dolomite and olivine on producing a low-sulfur pyrolysis oil. Energy. 2016 Nov 1;**114**:457-464

[33] Council Directive 2000/53/ EC. of 27 June 2002 on end-of-life vehicles.

Official Journal of the European Communities. L170: 81-82

[34] CONAMA. RESOLUÇÃO CONAMA 258. Brasília; 1999

[35] Li W, Wang Q , Jin J, Li S. A life cycle assessment case study of ground rubber production from scrap tires. International Journal of Life Cycle Assessment. 2014;**19**(11):1833-1842

[36] Sun X, Liu J, Hong J, Lu B. Life cycle assessment of Chinese radial passenger vehicle tire. International Journal of Life Cycle Assessment. 2016 Dec 1;**21**(12):1749-1758

[37] Pennington DW, Potting J, Finnveden G, Lindeijer E, Jolliet O, Rydberg T, et al. Life cycle assessment Part 2: Current impact assessment practice. Environment International. 2004;**30**:721-739

[38] Finnveden G, Hauschild MZ, Ekvall T, Guinée J, Heijungs R, et al. Recent developments in Life Cycle Assessment. Journal of Environmental Management. 2009;**91**(1):1-21

[39] International Organization for Standardization. ISO- TC 207/SC 5. ISO 14040:2006 Preview Environmental management. Life cycle assessment–Principles and framework. 2006. p. 20

[40] Guinée JB, Gorrée M, Heijungs R, Huppes G, Kleijn R, Koning, de A, et al. Handbook on life cycle assessment. Operational guide to the ISO standards. I: LCA in perspective. IIa: Guide. IIb: Operational annex. III: Scientific background. Dordrecht: Kluwer Academic Publishers; 2002. 692 p

[41] Hauschild MZ, Goedkoop M, Guinée J, Heijungs R, Huijbregts M, Jolliet O, et al. Identifying best existing practice for characterization modeling in life cycle impact assessment. International Journal of Life Cycle Assessment. 2013;**18**(3):683-697

[42] Kägi T, Dinkel F, Frischknecht R, Humbert S, Lindberg J, De Mester S, et al. Session "Midpoint, endpoint or single score for decision-making?"-SETAC Europe 25th Annual Meeting, May 5th, 2015. International Journal of Life Cycle Assessment. 2016;**21**(1):129-132

[43] Chiu C-T, Hsu T-H, Yang W-F. Life cycle assessment on using recycled materials for rehabilitating asphalt pavements. Resources, Conservation and Recycling. 2008;**52**:545-556

[44] Galvagno S, Casu S, Casabianca T, Calabrese A, Cornacchia G. Pyrolysis process for the treatment of scrap tyres: Preliminary experimental results. Waste Management. 2002 Dec;**22**(8):917-923

[45] Beukering PJH, Janssen MA. Trade and recycling of used tyres in Western and Eastern Europe. Resources, Conservation and Recycling. Vol. 33. 2001

[46] Goedkoop M, Spriensma R. The Eco-Indicator 99–A damage oriented method for Life Cycle Impact Assessment–Methodology Report. Amersfoort; 2001. Available from: https://www.pre-sustainability.com/download/EI99_annexe_v3.pdf

[47] Stindt D, Sahamie R. Review of research on closed loop supply chain management in the process industry. Flexible Services and Manufacturing Journal. 2014;**26**(1-2):268-293

[48] Karaağaç B, Ercan Kalkan M, Deniz V. End of life tyre management: Turkey case. Journal of Material Cycles and Waste Management. 2017;**19**(1):577-584

Chapter 5

Perspectives on Subnational Carbon and Climate Footprints: A Case Study of Southampton, UK

Laurence A. Wright, Ian D. Williams, Simon Kemp and Patrick E. Osborne

Abstract

Sub-national governments are increasingly interested in local-level climate change management. Carbon- (CO_2 and CH_4) and climate-footprints—(Kyoto Basket GHGs) (effectively single impact category LCA metrics, for global warming potential) provide an opportunity to develop models to facilitate effective mitigation. Three approaches are available for the footprinting of sub-national communities. Territorial-based approaches, which focus on production emissions within the geo-political boundaries, are useful for highlighting local emission sources but do not reflect the transboundary nature of sub-national community infrastructures. Transboundary approaches, which extend territorial footprints through the inclusion of key cross boundary flows of materials and energy, are more representative of community structures and processes but there are concerns regarding comparability between studies. The third option, consumption-based, considers global GHG emissions that result from final consumption (households, governments, and investment). Using a case study of Southampton, UK, this chapter develops the data and methods required for a sub-national territorial, transboundary, and consumption-based carbon and climate footprints. The results and implication of each footprinting perspective are discussed in the context of emerging international standards. The study clearly shows that the carbon footprint (CO_2 and CH_4 only) offers a low-cost, low-data, universal metric of anthropogenic GHG emission and subsequent management.

Keywords: urban metabolism, cities, community GHG, GHG inventory, carbon footprint

1. Introduction

Increasing GHG emissions have catalysed urban GHG management, with many having established sub-national and transnational climate networks, initiatives or management plans [1]. Carbon- (CO_2 and CH_4) and climate-footprints—(Kyoto Basket GHGs), are single impact category—global warming potential—indicators of life cycle assessment (LCA). These metrics provide an opportunity to develop effective models of GHG emissions from cities, and to facilitate effective mitigation. The frameworks required to calculate a carbon or climate footprint also provide a framework for the application of a more holistic LCA to cities or other geographic areas.

IntechOpen

Discussions to date have primarily focused on the appropriateness of the allocation of emissions to the local level, with progress driven by improved understanding of urban metabolism—material and energy flows through the urban system (e.g. [2–6]). Approaches can be categorised as process-led bottom-up approaches, top-down economic led analysis, or top-down "natural laboratory" approaches relying on atmospheric measurement and concentration [7].

"Territorial-based" (alternatively, "in-boundary", "geographically-based", or "production-based") approaches, generally adaptations of the *IPCC Guidelines for National Greenhouse Gas Inventories*, or the *Greenhouse Gas Protocol* developed for corporate GHG reporting, account emissions within geopolitical boundaries [8, 9]. These methods successfully identify local emissions patterns and inform local development policy. However, there has been growing recognition that holistic management of urban GHGs necessitates the inclusion of direct and indirect emissions as urban economies demand resources beyond their geographic locations [5, 10, 11].

"Transboundary" (alternatively, "territorial-plus", "geography plus" or "metabolism-based") approaches add out-of-boundary emissions associated with economic demand to territorial emissions, with the exact boundary conditions and scope varying between studies [2, 4–6]. Top-down "consumption-based" methods include all emissions along the supply-chain of goods and services, with boundary conditions defined by final consumption of households and governments [5]. This approach is useful in the informing mitigation of emissions associated with final consumption, although the exact origin of embodied emissions cannot normally be delineated and emissions from local production for exports are excluded [11]. Consequently, methods are not sensitive to many local strategies to reduce emissions [5].

Ultimately both concepts are complementary, focusing on different aspects of community composition . The primary cause of inconsistency between studies (for a review see: [3, 10]) and emerging standards (e.g. [12, 13]) is the approach taken to boundary conditions (spatial and temporal). Temporal boundaries vary, but typically consider an annual period, with some models operating at finer scales (e.g. [7, 14]). Spatial boundaries vary reflecting goals and application, and the lack of a singular definition for a city or an urban area. However, a 'city' or and 'urban area' is simply a taxonomic division of a 'community'—a specific area or place considered jointly with its inhabitants. Spatial boundaries can thus be decided on a case-by-case basis, defined by motivation application [1].

This chapter reviews urban sectorial methods, results, and policy implications of applying a territorial, transboundary, and consumption-based, carbon or climate footprint to a city, using a case study of Southampton, UK. Based on the framework proposed by Wright et al. [15] the requirements and methods to assess a carbon or climate footprint are presented. We disassemble the framework into 'modules', recognising that each element of the framework would require separate calculation methods. This enables the development of novel methods or the use of existing methods in a novel manner to create an overall methodology for the calculation of all elements of the framework. As proof of concept and to inform the development, the methodology was applied to Southampton, UK. Results are then presented for the carbon footprint and the 'climate footprint' ([15, 16], respectively). The methodology represents a novel approach, building on established practice to enable the sub-national assessment of carbon footprints in communities, which enables the spatial and temporal reporting of results at a sub-community level to enable effective management and policy development. We discuss the results and policy implications and conclude with a consideration of the effectiveness of current practice and highlight ongoing issues.

2. Case study: city of Southampton

Southampton (pop. 239,428 during study period), chosen as a case study as it contains the representative components of many cities, is the largest city in Hampshire, England (area: 51.91 km^2) based on the geographic extent of the city geo-political boundary [17]. The city is governed by Southampton City Council, a unitary authority (a single tier local government responsible for local government functions); the wider region is within the remit of Hampshire County and multiple district councils (a hierarchical system of governance common to many countries).

Southampton is a commerce hub; a major international cruise terminal, and the UK's second largest container port. A significant proportion of Southampton's workforce (circa 42%) commutes from the wider region and surrounding counties [18]. The city has two universities with a transitory student population of in excess of 40,000 [19]. Southampton Airport is a regional domestic and international airport located just outside the city's geopolitical boundary.

3. Methods

3.1 Residential

Large communities contain a significant number of dwellings, and emissions are driven by energy consumption, highly dependent on building structure and the behaviour of residents [20]. Estimation of emissions can be made on fuel consumed (e.g. sales) (e.g. [21]), however this method does not allow for spatial disaggregation. Alternatively energy use can be estimated from census based residential energy consumption models (e.g. [22]). Various methods have been developed for this purpose (for review see [23, 24]). To enable spatial disaggregation the case study applies the assumption that energy use can be simplified with the application of 'average' building categories. Model generalisation parameters are derived for categories of dwelling and applied to individual property build forms with Geographic Information Systems (GIS), eliminating the need for visual inspection [25]. Parameters were derived from the Building Research Establishment Domestic Energy Model (BREDEM) (BREDEM-8—monthly or BREDEM-12—annual) [26]. Total energy demand was assumed to be met using a combination of electricity, natural gas, and other fuels. Consumption data for electricity and natural gas were available from local metering records, with remaining demand assumed to be met using others fuels apportioned on basis of regional sales data. Output is restricted to an aggregation of properties rather than the individual building level, as accuracy would be open to significant variation and introduces confidentiality concerns.

3.2 Commercial and industrial point sources

The commerce and industry sector encompasses emissions associated with industrial physical or chemical processing and non-electrical energy. Complexities exist in the allocation of emissions between the energy and processing sectors (e.g. residual heat may be used for electricity generation). Actual consumption data from sales records or feedstock records is difficult to obtain, primarily due to the sensitive nature of such data. Point source data from larger facilities may be available from legislative emissions reporting schemes, although this often does not encompass small schemes. Proxy consumption data for fuels and processes can be utilised to estimate emissions, however this assumes fuel combustion at place of purchase, and may not accurately reflect the source of emissions. Gurney et al. [7] describe a model to simulate energy demand based on building parameters combined with known local atmospheric emissions. The same study notes that this method is only suitable for large point source emitters. Alternatively, pro-rata allocation of national emissions to local sources provides a reliable method of estimation (e.g. [7, 27]). For the purpose of the case study, supplemented with meter point natural gas and electricity consumption data, emissions by industry were pro-rated on employment by industrial sector (to a 4 digit Standard Industry Classification (SIC2007)).

3.3 Electricity, heat, and steam

Transboundary emissions relating to electricity are commonly calculated using an aggregated factor representing a national system of generators and transmission. Emissions from heat and steam are often reported separately due to data conventions and that composite emissions factors may over- or under- estimate of emission intensity [21]. Similarly, aggregated emissions do not segregate in-boundary generation, or consider low GHG decentralised generation schemes—likely to be a component of meeting carbon reduction targets [28]. In these cases electricity generated in-boundary and fed into national supply grids is representative of the grid average. Alternatively to provide greater disaggregation emissions associated with in-boundary electricity generation can be reported separately, either as a proportion of total consumed or as with the case study an absolute. Emissions for Southampton's electricity consumption were calculated using a national grid emissions factor (accounting for transmission, transformation and other losses (typically circa 6–11%) [21]), estimated from national generation and electricity consumption.

3.4 Road transport

Road transport emissions are often artificially truncated at the city boundary, but commuting represents a significant transboundary emissions source [11]. Economic data on fuel sales can be a viable indicator of road transport emissions, where the study area represents a commuter-shed [21]. However, this method is less effective where significant numbers of commuter trips occur (e.g. Southampton—circa 42% of work related trips are from outside the city [17]). In these cases the location of fuel purchase is not necessarily representative of fuel consumption. An alternative method is through the use of proxy relationships, with emissions estimated through regression based approaches [29] or population density and road density [30]. High temporal and spatial resolutions have been achieved using activity-based approaches, combining vehicle kilometres travelled (VKT) with fleet and fuel data [7, 31]. This approach requires total distance travelled by all vehicles in the study area, fuel efficiency, and fleet composition. Issues arise in comparability of VKT techniques as many cities have their own bespoke modes [21]. However this has the advantage of allowing bespoke modelling of spatial and temporal impacts of traffic policy intervention at high resolutions.

The basic principle of an activity based models is the relationship of the mass of fuel consumed in the distance travelled. The amount of fuel a vehicle consumes in a given distance is dependent on a number of parameters, including drive cycle, engine temperature, ambient temperature, fuel type, and fuel quality [32]. Hot-start emissions were calculated; by modal split, fuel type and installed vehicle technology, using experimentally derived emissions factors for vehicle type and pollutant by trip length and velocity from the ARTEMIS (Assessment and Reliability of Transport Emission Models and Inventory Systems) methodology and TRL emission factor database [32]. Cold start emissions are accounted using an excess factor over the hot-start emissions rate [33, 34].

3.5 Rail

Trips by rail transportation typically traverse the geopolitical boundary of a number of communities. Rail journeys involve a series of embarkation points between origin and destination, often with multiple stopping locations within the geopolitical boundary. A boundary limited methodology does not account the transboundary demand driven nature of these trips. For trips that originate outside the community boundary only the in-boundary proportion of the trip is accounted, conversely pass through trips that are not a result of city demand are still counted [4]. This issue exacerbates when considering national and internationally connected rail networks—a trip could begin a significant distance from the study community. Accounting in-boundary and transboundary emissions related to rail commuting creates the potential for double counting between communities. Pass through trips are accounted as a direct emission and then accounted again at the destination community. Reflecting these difficulties a number of community based GHG inventories do not explicitly define emissions from rail transportation (e.g. [4, 21]).

These issues can be addressed by accounting emissions based on proportional commuter distances travelled. Assigning emissions from rail commuter demand as passenger kilometres travelled to total passenger kilometres travelled on the relevant routes offers a mechanism to apportion trips to the local community a demand basis. Accounting both in-boundary and transboundary emissions requires a combination of two methods—one to calculate in-boundary emissions and another to allocate transboundary demand emissions. For the purpose of the case study, in-boundary emissions were calculated using ARTEMIS technology specific bottom-up algorithms and emissions factors (function of engine, technology, distance and speed) [32]. All journeys on non-electrified rail were assumed to be power by diesel. Trips on electrified rail were apportioned to diesel or electric locomotives using operator timetables. Total trips, distance travelled, and operational engine time were estimated from train operator time tables [35], combined with the Ordnance Survey Integrated Transport Layer [36]. Emissions associated with commuter trips were estimated as a function of rail demand for Southampton, passenger kilometres travelled [37] were estimated as proportional to the total ticketed exits on the national rail network (collected by automated barrier passes) divided by number of ticketed exits at Southampton.

3.6 Other off-road mobile emissions

Mobile off-road sources represent an extremely diverse range of domestic and commercial emissions. Including controlled activities which are consistent and follow specific procedures (e.g. dockside grab loader) and chaotic activities following no pre-determined procedures or activity patterns (e.g. domestic lawn mowers) [38]. Fuel sales data may be a viable indicator of emissions, where the operation of off-road machinery are geographically constrained to the location of fuel purchase [39], although this method fails where fuel purchase does not represent the location of consumption.

Unlike road transport, the majority of off-road machinery units are not registered making estimation of populations and activity difficult. Proxy estimates of population can be made based on national purchases or populations pro-rated to the local level, as per the case study [40]. This assumes a uniform distribution of machinery across total national population, which may not be representative of local conditions. Alternative allocation methods could be utilised that consider a number of machinery units as a function of purpose or spatial area (e.g. lawnmowers f(greenspace), construction machinery f(growth)), however the wide range and chaotic usage patterns of off-road machinery are likely to confuse this issue.

3.7 Shipping

Cities that are international cruise and container terminals rely heavily on these industries for economic growth and employment, exclusion of emissions from these industries would lead to misinterpretation in policy making [1]. Territorial inventories may, depending on the extent of territorial waters in the geopolitical boundary, include port-side operations or entirely exclude shipping operations. A transboundary approach must consider the indirect emissions (movement between ports) of these sources [1, 5]. Emissions from shipping are a function of engine operation and fuel consumption. Calculation of fuel consumed has broadly been undertaken using two approaches—'engine use models' and 'bunker fuels'. Engine use models apply engine load, power and run-time, by engine and ship type (e.g. ro-ro ferry, liquid bulk), in the three phases of operation (hoteling, manoeuvre and cruise) to calculate emissions [41]. This requires detailed data input on vessel characteristics, routes, and operational time. Detailed data for all ship movements (>250 gross tons) and characteristics are available from the from historic Automatic Identification System datasets. However, the majority of these datasets demand a high cost purchase, which excludes some sub-national governments from using the data (e.g. Lloyds List Intelligence [42]). Alternatively, the method taken in the case study, a bunker fuels approach considers international bunker fuels loaded at the departure port provide a proxy to estimate emissions from shipping [41]. However, shipping companies are likely to source the cheapest available fuel for the route, the result being where fuel cost is low, emissions are overestimated (e.g. Belgium), and where costs are high, emissions are underestimated (e.g. New Zealand) [43, 44].

3.8 Aviation

Aviation emissions are transboundary, smany airports are located outside geopolitical boundaries, and cities often act as aviation hubs with transit passengers occupying a significant proportion of capacity [45]. Allocation of emissions must address these concerns, so as not to generate political tensions. Some territorial studies exclude emissions as almost entirely transboundary and largely beyond the control of local government (e.g. [46–48]). Others include domestic emissions and take-off and landing cycles to 1000 m altitude for international emissions (e.g. [49]). As applied in the case study emissions can be calculated on an activity basis (engine runtime, technology, flight occupancy). Similarly a number of studies have reported transboundary emissions based on quantities of fuels loaded at airports within city boundaries (e.g. [14]). These methods do not consider the movement of passengers between flights and the surface movement of passengers from outside city limits. Previous authors suggest that regional airport usage by community inhabitants can be estimated as a function of local to regional population [4, 21]. Assignment of emissions by community demand offers a truer picture emissions, considering only those emissions associated with the local population. However, this method is fraught with complexity, especially in cases where a number of international airports operate within close proximity (e.g. southern UK—Southampton; Bournemouth; Gatwick; Heathrow; Stansted, London City). Without accurate passenger origin—destination data, subjective judgments must be made to establish the geographic extent of airport demand. Demand from beyond the geographic boundary could be considered a function of the community demand, thus arguably, related aviation emissions should be accounted [4].

3.9 Agriculture, Forestry and Other Land Use (AFOLU)

Some argue AFOLU is potentially insignificant at the urban level and therefore may be excluded [50]. This is based upon the assumption that green space is both relatively limited in urban centres, and the perception that urban green space has limited value due to human modification [51, 52]. This is often untrue (e.g. Southampton Common is 145 hectares; London's Hyde Park is 142 hectares, Beijing's Fragrant Hills Park is 160 hectares, and Vancouver's Stanley Park is >400 hectares), and fails to consider the importance of public and private land in urban centres (e.g. private gardens, green roofs) which, whilst small compared to per unit area GHG emissions, are potentially important stocks of GHGs [53].

Land use and management significantly influences ecosystem processes that effect GHG fluxes, (e.g. photosynthesis, respiration, decomposition). The IPCC [8] guidelines for national inventories contain significant information for the calculation of AFOLU GHG fluxes. These guidelines suggest two methods: (i) net carbon stock change over time and (ii) direct carbon flux rate (more commonly utilised for non-CO_2 species) [8]. AFOLU carbon flux for Southampton was calculated using the first option, to provide consistency with annual reports and promote favourable management of non-urbanised space over an extended time scale.

Estimates of C flux were derived from Rothamsted soil carbon model (RothC-26.3) and the Lund-Potsdam-Jena Dynamic Global Vegetation Model (LPJ-DGVM) [54–56]. Basic climatic inputs (temperature, precipitation, daylight hours) were required (Met Office, 2014), in addition to data on organic matter inputs (obtained from LPJ-DGVM), soil clay content, and atmospheric CO_2 concentrations [56]. GIS data (OS MasterMap) of land-cover types were used to create a map of the city area; where available this map was augmented with specific vegetation cover data provided by the municipal authority [36]. Land-cover data was classified into 11 broad categories, adapted from a condensed set of JNCC Phase 1 habitat classifications, a standard mode of habitat classification in the UK. (**Table 1**) The Phase 1 habitat classifications provide a specific name and brief description of each habitat type/feature, appropriate for vegetation modelling using LPJ-DGVM [54–57]. In cases where land-cover types are not complete for an area (e.g. scattered trees), the land-cover was assumed to be divided evenly between land-cover types. Where trees are described as 'scattered' (>30% of surface by canopy extent) 20% of total area is classified as that tree type, the remainder is divided evenly between other represented land-cover types [57]. In the grass (cut) category, data are required for total clippings collected, thus removed from the system, and total clippings left in-situ.

Private gardens are representative of multiple land-cover types (e.g. lawn; ornamental planting; patios; tarmac; gravels). Typical land cover types in private gardens were estimated based on a representative sample of private gardens in the study area, categorised for land cover types using aerial photography (expert judgement) (**Table 2**).

The model was run across a temporal period of 1 year, with GHG flux calculated as the change in storage between runs.

3.10 Waste

Waste management generates emissions of CO_2, primarily of biogenic origin, with some fossil carbon and CH_4, often outside the city boundary [5]. Regional or municipal governments are both actors in and managers of waste. Each has their own waste infrastructure, service provision and socio-economic conditions with influence over collection; treatment, and destination with significant emission

Land-cover category	Example land-cover types
Grass (cut 11 times a year)	Natural surface, slope
Rough grass (not mown)	Rough grass, rough grass and other
Other herbaceous plants	Perennials, flowers, roses
Private gardens	Multiple surfaces in private residence
Broadleaved summergreen trees	Non-coniferous trees, scattered non-coniferous trees, orchard
Needle leaved evergreen trees	Coniferous trees, scattered coniferous trees
Scrub	Scrub, shrubs, hedges, heath
Marsh	Marsh reeds or saltmarsh
Sealed surface	Road, made surface, paths, steps, track, structure, traffic calming, pylon, rail, upper level of communications, building, glasshouse, overhead construction, unclassified
Water	Inland water, foreshore, tidal water

Table 1.
Land-cover categories for modelling of vegetation or other land-cover types (adapted from [49]).

Land-cover type	Proportion of total area (%)
Grass (cut 11 times a year)	
Clippings removed	10
Clippings left in situ	30
Shrubs	10
Temperate broadleaved summergreen trees	10
Other herbaceous plants	10
Sealed	30

Table 2.
Assumed proportions of land-cover types in private gardens for the southern UK (expert judgement).

savings available through system reconfiguration [58, 59]. Many previous studies apply 'generic' emissions factors to waste treated. Detailed tools and methods for the accounting of GHG emissions from waste systems have been developed although there are concerns regarding consistency, accuracy and transferability of these methods [60, 61]. The following offers a brief overview of methods applied in the case study with greater detail exploring various stages in the waste system given in supplementary information.

Knowledge of waste composition and subsequent mass balance of CO_2 and CH_4 throughout the waste system is the key determinate in modelling waste emissions. The composition of the wastes in the treatment system will affect the mass balance due to the changes in carbon content and subsequent degradation patterns [60]. Once known waste stream emissions can be calculated on a mass balance or activity basis. Many city based assessments do not suggest breakdown of the various stages of the waste management process, instead offering per unit treated emissions factors. Per unit emissions factors are applicable to processes where the primary main source of emissions are from energy use (e.g. waste recovery and recycling), or for incineration processes (mass balance of carbon could also be applied). Greater accuracy can be achieved in modelling biological processes using a mass balance approach, with CO_2 and CH_4 emissions calculated on mass balance of carbon input to carbon lost from the final product. Emissions estimates from landfill must recognise both operational and closed phases [62]. Following closure, a landfill continues to emit GHGs, possibly for several hundred years, although some carbon will be indefinitely stored in the landfill [63]. Kennedy et al. [21] propose a pragmatic solution, applied in the case study, whereby estimates of long-term emissions were calculated for the waste landfilled in the assessment year.

3.11 Water

The provision of water and waste water services are similar to the provision of electricity. Emissions associated with water are calculated on an end-user basis for water processing, treatment and transportation using per unit consumed emissions factors. Commonly, as during this case study, water use is not metered and thus no actual consumption data are available. In the UK significant effort is being directed to the installation of end-user metering; this will provide improved data resolution for future investigations [64]. Emissions were calculated using standard estimates of water consumption provided by water suppliers.

3.12 Consumption

It is generally accepted that the addition of a consumption-based modelling approach extends the research implications and policy potential of a GHG inventory [2]. Territorial accounts include emissions associated with exports at the point of production; but exclude those associated with supply chains and imports. The upstream impacts of production are allocated to the producer—the tendency is to mask embedded emissions and burden shifting (energy intensive industries are effectively exported). Transboundary approaches add an element of these out-of-boundary emissions, but do not give a full picture of the impact of consumption. Consumption-based accounting focuses on the final consumption of households and governments; methods account all GHG emissions upstream of the community but exclude emissions from production within the city [10]. A consumption based approach compliments a transboundary methodology, capturing emission flows and the driving forces associated with consumption [65].

A consumption-based approach requires linking supply chain emissions with local consumption activities. Input-output (IO) models detail the transactions between industries and sectors within the economy. An IOT requires knowledge of all flows of goods and services among intermediate and final sectors in disaggregated form for a given time period. This implicitly implies high volumes of data, which is difficult to obtain at the sub-national level, necessitating some form of scaling from national data. The core element of an input-output model is a matrix concerning flows through the economy—sales and purchases from an industrial sector (a producer), to other sectors and the sector itself (consumers) [66]. The basic input-output out model assumes homogeneity in sectors (i.e. each sector produces a single product) and linear production (i.e. proportionality of inputs and outputs which precludes economies of scale). The basic IO model can be extended to include material consumption and emissions—an Environmentally Extended

Input-Output (EEIO) model. Effectively this creates an 'environment' sector, and the value of each item represents the 'output' of pollution [67].

Consumption-based emissions factors (GHG/£ spent) for the UK were calculated using an EEIO model. The IO data only holds data on final consumption at the national level. A downscaling methodology was therefore required to estimate final consumption at the local level. The model assumes no variation in emissions per monetary unit spent between the national and local levels. The technical coefficient matrix was derived from UK supply and use tables for the year 2008 with 123 products and industry sectors in basic prices [68]. GHG emissions data by industry sector for the period were taken from the UK Environmental Accounts [69]. The Environmental Accounts provide data on GHG emissions from 129 industrial sectors and 2 household emissions sources (travel and non-travel). The GHG data is provided at a more disaggregated level than the IO data in some sectors, this was scaled to the 123 sectors of the IO model using the parent sector of the lower level disaggregation according to the UK Standard Industry Classification 2007. A domestic technology assumption is applied to imported goods and services, whereby imports are assumed to have the same GHG intensity as domestic equivalents. It assumes the energy structure and economic structure of the imports can be approximated based on the domestic make-up of the UK. This may be a valid assumption for some regions, but underestimates GHG intensities of imports from emerging and developing regions [11].

Expenditure between regions will vary considerably as a result of a range of socio-demographic factors. However, the underlying IO data only provide expenditure at the national level. Household demand was downscaled to the local level using household expenditure data from the UK Living Costs and Food Survey (LCF) (annual survey of household expenditure on consumer products and services), and derived summary datasets provided in the Family Spending report [70, 71]. Government expenditure was downscaled on a per capita basis. Whilst this assumes individuals in the national population benefit equally from all government expenditure, it is considered a reasonable assumption in the absence of alternative data. Researchers have downscaled government expenditure using local expenditure statistics, however these data do not exist for the UK [11]. The study does not consider emissions relating to capital investments.

4. Uncertainty

The city system is inherently complex and comprised largely of non-deterministic features (i.e. responses of the system that are not predicable because of uncertainty within the system itself). Qualification and assessment of these uncertainties is important for both model validation and reliability. Sensitivity analysis is used to assign the uncertainty in the output of the model to different sources of uncertainty in the model's inputs and how the model responds to changes in input data [72]. The sensitivity of the transboundary inventory model is considered using a one-at-time (OAT) local sensitivity analysis technique. Sensitivities for the consumption estimates are considered at the aggregated emission factor per unit expenditure level, rather than at the EEIO input variable level due to complexities involved in this form of modelling [73]. Whilst sensitivity analysis provides a good indicator of variables with high impact on the model, it does not provide qualification of uncertainty and must be accompanied with an uncertainty analysis [72]. A Monte Carlo analysis was performed using random sampling of input variables, based on defined uncertainty probability distributions in input parameters. The analysis consisted of ten thousand model runs, completed for the model as a whole and for three of the broad category areas identified in the OAT transport emissions; power generation; and waste disposal. Supply chain and consumption emissions uncertainty was excluded due a need for further investigation and modelling.

5. Results and discussion

5.1 Summary

As identified, there are three methods for the assessment of life-cycle GHG emissions from cities and other communities—territorial, transboundary, and consumption based. This section discusses the implication of the three methods using the Southampton case study. Furthermore, the Carbon Footprint (CO_2 and CH_4) and Climate Footprint (Kyoto Basket) metrics are compared for each method. The summary results (**Table 3**) indicate increasing size in both the carbon and climate footprints as further emissions sources are added between methods, and a slight increase between the carbon and climate footprint metric.

5.2 Territorial emissions

Southampton territorial emissions suggest carbon and climate footprints of 268ktCO$_2$e and 273ktCO$_2$e, respectively. Addition of end-use electricity consumption increases this figure by 601ktCO$_2$e and 604ktCO$_2$e, respectively (**Figure 1**). The minor increase (0.99%) in emissions between the total carbon and climate footprints is driven primarily through inclusion of additional GHGs in transport (primarily N_2O). Calculation of per capita emissions for the case study indicates 3.7 tCO$_2$e/capita carbon footprint, lower than the equivalent national production-based 10.32tCO$_2$e per capita estimate for the UK [74]. Whilst strictly geographic based methods can successfully identify local production-based emissions patterns and inform local development policy, they fail to capture the full extent of sub-national community infrastructures which extend beyond the geopolitical boundary (e.g. transport) [5, 6].

5.3 Transboundary emissions

Described by Ramaswami et al. [4], Denver (CO, USA) represents the first known community to have been inventoried using a transboundary methodology. The study accounted all in boundary emissions and identified key community flows defined as: food; water; transport, and building materials (for shelter). Hillman and Ramaswami [75] suggest, based on a study of eight US cities that these

Method	Carbon footprint (ktCO$_2$e)	Climate footprint (ktCO$_2$e)
Territorial	268	273
Territorial+	601	604
Transboundary	2643	2787
Consumption	3160	3590

Note: territorial+ includes emissions from end-user electricity consumption.

Table 3.
Summary carbon and climate footprints for the case study of Southampton.

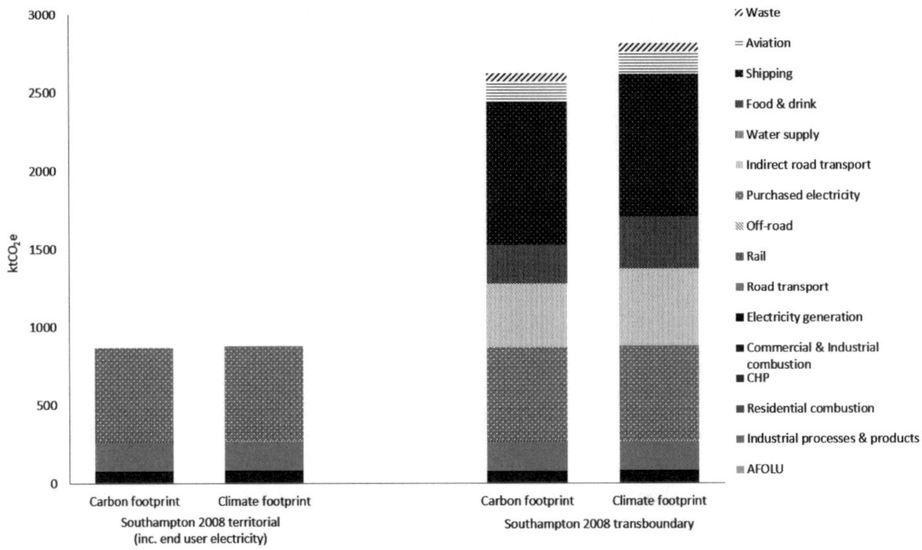

Figure 1.
Comparison of case study territorial and transboundary carbon (CO_2, CH_4) and climate (Kyoto Basket) footprints.

cross-boundary activities contribute on average 47% more than the in-boundary emissions sources. This consideration is reflected in developing international standards (e.g. [12, 13]) which suggest a transboundary approach to account both the territorial and transboundary aspects of a community—ideally moving towards an approach that replicates the process(s) of urban metabolism [2].

The Southampton transboundary inventory includes direct emissions with the addition of: commuter road transport; shipping; aviation; out-of-boundary waste emissions; water and wastewater supply/treatment; construction materials, and food and drink—representative of the requirements of recent PAS2070 standard. The 2008 results, carbon footprint 2643 ktCO_2e, and climate footprint 2787 ktCO_2e, are, as expected, substantiality larger than the comparative territorial results (**Figure 1**). The results of the Monte Carlo simulation suggest a 95% confidence interval of 3395–4295 ktCO_2e. The two footprinting techniques, as per territorial emissions methods produce results within 1%. The increased emissions in the climate footprint stem primarily from transboundary transport. The largest contributor, shipping emissions, are a result of the extended travel distance and subsequent high fuel demands. Whilst sub-national governments have limited control (typically only port-side operations) over these emissions sources, inclusion is important due to the strong economic reliance on these industries [1]. However sub-national governments do have access to control to address these emissions through local air quality control. Similarly road transport control can be found through air quality control and additional controls in planning and road management.

Energy emissions comprise a large component of total emissions, electricity provides the dominant contribution to this sector. The disaggregation of emissions related to heat production from emissions associated with electricity generation impacted >1% on emissions per unit electricity consumed. At a local level, renewables account for an equivalent grid emission of 3 ktCO_2e. Evidently emissions from electricity are mainly dependent on the intensity of supply, highlighting a powerful interlink between local and national policy making. This interlink will become particularly pertinent with the potential advent of locally led energy initiatives (e.g. micro-generation; rail electrification; electric vehicle charging networks) [76].

Emissions from AFLOU are minimal, however this masks the carbon stored in urban green space (470.00 ktCO_2e). Exclusion of these emissions assumes green space storage is minimal; the results demonstrate this may not be the case. Careful consideration must be given to development that affects community green space (both negative—e.g. green space urbanisation—and positive—e.g. installation of green roofs), for the creation of carbon sinks, the wider potential social, and economic benefits [52].

Supply chain and infrastructure related emissions form the majority of total transboundary emissions, highlighting the importance of supply chains in community footprinting. The recent PAS2070 [13] suggests further inclusion of all materials making >2% material contribution to the community. This would add a further 1315 ktCO_2e and 1435 ktCO_2e (carbon and climate footprint, respectively) to the Southampton results. However, there are concerns about double counting with the territorial element of the assessment.

The primary advantage of a transboundary footprint is the level of completeness created through inclusion of in-boundary emissions sources and transboundary infrastructures that supply these activities. Given this completeness, transboundary based footprints can be utilised to inform a broad range of mitigation and management strategies at the local, regional, and national scales. Additionally transboundary footprints are more relevant and easier to communicate to residents due to the inclusion of major activities included in personal and home carbon calculators [6].

The main shortcoming of the transboundary method is the inconsistency in approach and application of metrics between studies. Standards (e.g. PAS2070 [13], GHG Protocol for Community Reporting [12]) are emerging that attempt to clarify and develop consistency in reporting structures. Comparability is also difficult; results require normalisation to enable inter-community comparisons. The majority of territorial inventories are normalised using a per capita metric, however this may not be appropriate for transboundary approaches. Metrics for the representation and comparison of transboundary approaches require further research.

5.4 Consumption emissions

Results for Southampton (carbon footprint 3160 ktCO_2e, climate footprint 3590 ktCO_2e) (**Figure 2**) are consistent with previous studies where consumption-based estimates are higher than production-based emissions, with the majority of emissions driven by households [77]. The disparity between the carbon and climate footprint is higher (circa 12%), this is primarily driven by high emissions of N_2O in agriculture, highlighting the need for a climate footprint approach in certain situations where high emission of GHGs other than CO_2 and CH_4 occur [15].

The addition of a consumption based account extends the policy implications of a local GHG inventory [2]. The approach provides value for the assessment of household consumer lifestyle on GHG emissions, making the consumption impact of households and government visible [6]. Arguably a consumption based approach provides for the most rigorous method for per capita GHG comparison, as consumption is driven by the residents of a community. Additionally a consumption based approach can inform local policy to reduce supply chain emissions as, when accurate local data are available, imports/exports can be traced. Recognising these advantageous policy implications, the new PAS2070 requires the separate completion of both a transboundary inventory, and a consumption-based inventory [13]. However consumption-based methods are data intensive, and are only truly valuable where accurate IO data are available. Misallocation of emissions can occur where physical flows do not match monetary flows represented in local IO tables [6]. Additionally, the consumption method effectively divides the community into

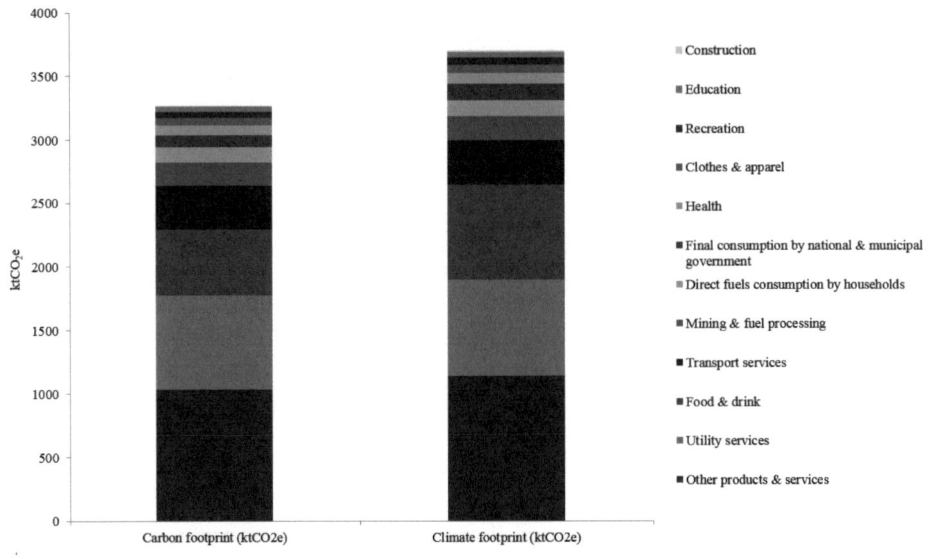

Figure 2.
Comparison of case study consumption carbon (CO_2, CH_4) and climate (Kyoto Basket) footprints.

two, with activities for exports not included in the unit of analysis. This can exclude a large element of a local economy (e.g. resorts, industrial communities) which could be managed through local policy.

In this study, there are limitations to note. The assumption of a homogenous technology mix in the EEIO model presents a level of inherent uncertainty—imports come from a range of countries using a range of different emission and resource intensities. This may be a valid assumption for some regions, but underestimates GHG intensities of imports from emerging and developing regions. The accepted solution is to employ a Multi-Region Input-Output (MRIO) model. MRIO models represent the interactions between any number of regions with potentially differing technology mixes, by internalising trade flows within internal demand [78]. The method of downscaling presents two important limitations. Firstly expenditure can only be estimated for broad categories—partially a result of the homogeneity assumption of the underlying IO model, this assumes common per unit emissions in these categories, which may not be entirely representative. Secondly, this generalisation may misrepresent the quantity of product purchased. For example the same expenditure on a high cost product variant would provide less quantity of product and potentially lower emissions, than a high quantity low cost product.

6. Conclusions

This study has presented several important developments to the assessment of community carbon footprints. Methods have been developed to assess emissions at a spatial and temporal disaggregation suitable for use by policy makers at the community level. The methods have been presented to show the policy implications of territorial, transboundary, and consumption based accounting procedures. To explore the uncertainties associated with the model a Monte Carlo simulation was constructed. The effort required for a comprehensive uncertainty analysis of this type is considerable, the alternative however, is to provide decision makers with incomplete information. At best this will lead to a false sense of reliability, at worse incorrect assumptions and decision making. We strongly recommend that as more studies become available continuous effort to identify and improve uncertainty be applied; leading to a better communication of information to policy makers and a better underpinning of their decisions. Only a limited difference in emissions totals was observed between the carbon and climate footprints for the case study city, clearly showing that the carbon footprint (CO_2 and CH_4 only) offers a low cost, low data, universal metric of anthropogenic GHG emission and subsequent management.

Territorial accounts may be suitable for national GHG inventories, but cannot represent the transboundary infrastructures of sub-national communities. Transboundary approaches extend the territorial approach to include emissions from key infrastructures essentially to sub-national communities. The addition of a consumption-based account further extends the policy relevance and research applications of community accounting. Consumption-based approaches show the impact of household consumer lifestyle on GHG emissions, and making the supply chain impact of households and government visible.

Recognising the advantages of transboundary and the simultaneous application of a consumption-based approach, standards, such as PAS2070, advocate combining a transboundary approach with a consumption-based approach in order to provide a comprehensive report.

Finally, the establishment of a global network of low carbon cities requires the appropriate tools. PAS2070 and related standards represent a significant step towards the development of a comparative assessment of urban community GHGs. Barriers still exist—comparable metrics need to be further developed and local governments often do not possess the resources and skills required to complete an inventory assessment.

Conflict of interest

The authors have no conflict of interest.

Author details

Laurence A. Wright[1], Ian D. Williams[2]*, Simon Kemp[3] and Patrick E. Osborne[3]

1 Warsash School of Maritime Science and Engineering, Solent University, UK

2 Faculty of Engineering and Physical Sciences, University of Southampton, UK

3 Faculty of Environmental and Life Sciences, University of Southampton, UK

*Address all correspondence to: idw@soton.ac.uk

References

[1] Wright LA, Williams I, Kemp S. The role of carbon footprinting in the development of global 'low-carbon' cities. In: Proc. LCA XI Conf.; Chicago, USA. 2012

[2] Baynes T, Weidmann T. General approaches for assessing urban environmental sustainability. Current Opinion in Environmental Sustainability. 2012;**4**:458-464

[3] Kennedy CA, Ramaswami A, Carney S, Dhakal S. Greenhouse gas emission baselines for global cities and metropolitan regions. In: Hoornweg D et al., editors. Cities and Climate Change. Washington D.C: The World Bank; 2011

[4] Ramaswami A, Hillman T, Janson B, Reiner M, Thomas G. A demand-centered, hybrid life-cycle methodology for city-scale greenhouse gas inventories. Environmental Science & Technology. 2008;**42**(17):6455-6461

[5] Wright LA, Coello J, Kemp S, Williams ID. Carbon footprinting for climate change management in cities. Carbon Management. 2011;**2**(1):49-60

[6] Chavez A, Ramaswami A. Articulating a trans-boundary infrastructure supply chain greenhouse gas emission footprint for cities: Mathematical relationships and policy relevance. Energy Policy. 2012;**54**:376-384

[7] Gurney KR, Razlivanov I, Song Y, Zhou Y, Benes B, Abdul-Massih M. Quantification of fossil fuel CO_2 emissions on the building/street scale for a large U.S. City. Environmental Science & Technology. 2012;**46**(21):12194-12202

[8] IPCC. 2006 IPCC guidelines for national greenhouse gas inventories, prepared by the national greenhouse gas inventories programme. In: Eggleston HS, Buendia L, Miwa K, Ngara T, Tanabe K, editors. Japan: IGES; 2006

[9] Ranganathan J, Corbier L, Bhatia P, Schmitz S, Gage P, Oren K. The Greenhouse Gas Protocol: A Corporate Accounting and Reporting Standard (Revised Edition). Geneva, Switzerland: World Business Council for Sustainable Development and World Resources Institute; 2004

[10] Chavez A, Ramaswami A. Progress toward low carbon cities: Approaches for transboundary GHG emissions' footprinting. Carbon Management. 2011;**2**(4):471-482

[11] Minx J, Baiocchi G, Wiedmann T, Barrett J, Creutzig F, Feng K, et al. Carbon footprints of cities and other human settlements in the UK. Environmental Research Letters. 2013;**8**(3):1-10

[12] Arikan Y, Desai R, Bhatia P, Fong WK. Global Protocol for Community-Scale Greenhouse Gas Emissions (GPC) Pilot Version 1.0, C40 Cities Climate Leadership Group and ICLEI Local Governments for Sustainability in collaboration with: World Resources Institute, world Bank, UNEP, and UN-HABITAT. 2012

[13] BSI. PAS2070: Specification for the Assessment of Greenhouse Gas Emissions of a City by Direct Plus Supply Chain, and Consumption-based Approaches. London: BSI; 2013

[14] Kennedy C, Steinberger J, Gasson B, Hansen Y, Hillman T, Havranek M, et al. Greenhouse gas emissions from global cities. Environmental Science & Technology. 2009;**43**(19):7297-7302

[15] Wright LA, Kemp S, Williams I. Carbon Footprinting: Towards a universally accepted definition. Carbon Management. 2011;**2**(1):61-72

[16] Williams I, Kemp S, Coello J, Turner DA, Wright LA. A beginner's guide to carbon footprinting. Carbon Management. 2012;**3**(1):55-67

[17] Office for National Statistics (ONS). Population Estimates for UK, England and Wales, Scotland and Northern Ireland, Mid 2008 [data file] Crown Copyright, London. 2010

[18] Department for Transport (DFT). National Travel Survey. London, UK: Department for Transport (DFT); 2008

[19] Zhang N, Williams I, Kemp S, Smith N. Greening academia: Developing sustainable waste management at higher education institutions. Waste Management. 2011;**31**(7):1606-1616

[20] Firth SK, Lomas KJ, Wright AJ. Targeting household energy-efficiency measures using sensitivity analysis. Building Research & Information. 2010;**38**(1):25-41

[21] Kennedy C, Steinberger J, Gasson B, Hansen Y, Hillman T, Havránek M, et al. Methodology for inventorying greenhouse gas emissions from global cities. Energy Policy. 2010;**38**(9):4828-4837

[22] Zhou Y, Gurney K. A new methodology for quantifying on-site residential and commercial fossil fuel CO_2 emissions at the building spatial scale and hourly time scale. Carbon Management. 2010;**1**(1):45-56

[23] Kavgic M, Mavrogianni A, Mumovic D, Summerfield A, Stevanovic Z, Djurovic-Petrovic M. A review of bottom-up building stock models for energy consumption in the residential sector. Building & Environment. 2010;**45**(7):1683-1697

[24] Natarajan S, Padget J, Elliott L. Modelling UK domestic energy and carbon emissions: An agent-based approach. Energy and Buildings. 2011;**43**(10):2602-2612

[25] Rylatt RM, Gadsden SJ, Lomas KJ. Methods of predicting urban domestic energy demand with reduced datasets: A review and a new GIS-based approach. Building Services Engineering Research and Technology. 2003;**24**(2):93-102

[26] Anderson BR, Chapman PF, Cutland NG, Dickson CM, Doran SM, Henderson G. BREDEM-8: Model Description—BR439. UK: Building Research Establishment (BRE); 2002

[27] Tsagataskis ITB, Passant N, Brophy N. UK Emission Mapping Methodology 2008, AEA. Oxford, UK: Department of Environment, Food & Rural Affairs (Defra); 2008

[28] Allen SR, Hammond GP, Mcmanus MC. Prospects for and barriers to domestic micro-generation: A United Kingdom perspective. Applied Energy. 2008;**85**(6):528-544

[29] Brondfield MN, Hutyra CK, Gately SM, Raciti SM, Peterson SA. Modelling and validation of on-road CO_2 emissions inventories at the urban regional scale. Environmental Pollution. 2012;**170**:113-123

[30] Shu Y, Lam NSM. Spatial disaggregation of carbon dioxide emissions from road traffic based on multiple linear regression model. Atmospheric Environment. 2011;**45**:634-640

[31] Gately CK, Hutyra I, Wing S, Brondfield MN. A bottom-up approach to on-road CO_2 emissions estimates: Improved spatial accuracy and applications for regional planning. Environmental Science & Technology. 2013;**47**(5):2423-2430

[32] Boulter PG, McCrae IS. ARTEMIS: Assessment and Reliability of Transport Emission Models and Inventory Systems—Final Report; TRL. Wokingham, UK: TRL; 2007

[33] Joumard R. Methods of estimation of atmospheric emissions from transport: European scientist network and scientific state-of-the-art, INRETS report LTE, 9901. 1999

[34] Nejadkoorki N, Nicholson K, Lake I, Davies T. An approach for modelling CO_2 emissions from road traffic in urban areas. Science of the Total Environment. 2008;**406**:269-278

[35] Network Rail. The Working Timetable (WTT). London, UK: Network Rail; 2010

[36] Ordnance Survey. MasterMap Integrated Transport Layer. Southampton, UK: Crown Copyright; 2014

[37] Office of Rail Regulation (ORR). Passenger Kilometres by Year [Data File]. London, UK: Office of Rail Regulation; 2014

[38] Lewis P, Rasdorf W, Frey HC, Pang SH, Kim K. Requirements and incentives for reducing construction vehicle emissions and comparison of nonroad diesel engine emissions data sources. Journal of Construction Engineering and Management. 2009;**135**(5):341-351

[39] Colodner S, Mullen MA, Salhotra M, Schreiber J, Spivey M, Thesing KB, et al. Port authority of New York and New Jersey criterion pollutant and greenhouse gas emission inventory. Transportation Research Record: Journal of the Transportation Research Board. 2011;**2233**(1):53-62

[40] Assessment and Standards Division EPA. Median Life, Annual Activity, and Load Factor Values for Non-road Engine Emissions Modeling nr-005d. Washington, DC, USA: EPA Office of Transportation and Air Quality; 2010

[41] EEA (European Environment Agency). EMEP/CORINAIR Emission Inventory Guidebook. Copenhagen: EEA; 2009

[42] Informa Plc. Lloyds List Intelligence. London, UK: Informa Plc; 2014

[43] De Meyer P, Maes F, Volckaert A. Emissions from international shipping in the Belgian part of the North Sea and the Belgian seaports. Atmospheric Environment. 2008;**42**(1):196-206

[44] Fitzgerald WB, Howitt OJ, Smith IJ. Greenhouse gas emissions from the international maritime transport of New Zealand's imports and exports. Energy Policy. 2011;**39**(3):1521-1531

[45] Harris PG, Chow ASY, Symons J. Greenhouse gas emissions from cities and regions: International implications revealed by Hong Kong. Energy Policy. 2012;**44**:416-424

[46] Brown MA, Southworth F, Sarzynski A. The geography of metropolitan carbon footprints. Policy and Society. 2009;**27**(4):285-304

[47] Parshall L, Gurney K, Hammer SA, Mendoza D, Zhou Y, Geethakumar S. Modeling energy consumption and CO_2 emissions at the urban scale: Methodological challenges and insights from the United States. Energy Policy. 2010;**38**(9):4765-4782

[48] Sovacool BK, Brown MA. Twelve metropolitan carbon footprints: A preliminary comparative global assessment. Energy Policy. 2010;**38**(9):4856-4869

[49] Bush T, Tsagtakis I, King K, Passant N. NAEI UK Emission Mapping Methodology 2006. Oxford: AEAT; 2008

[50] United Nations Environment Program (UNEP). International Standard for Determining Greenhouse Gas Emissions for Cities. Nairobi, Kenya: UNEP; 2010

[51] Tratalos J, Fuller RA, Warren PH, Davies RG, Gaston KJ. Urban form,

biodiversity potential and ecosystem service. Landscape & Urban Planning. 2007;**83**:308-317

[52] Davies ZG, Edmondson JL, Heinemeyer A, Leake JR, Gaston KJ. Mapping an urban ecosystem service: Quantifying above-ground carbon storage at a city-wide scale. Journal of Applied Ecology. 2011;**48**(5):1125-1134

[53] Pataki DE, Alig RJ, Fung AS, Golubiewski NE, Kennedy CA, McPherson EG, et al. Urban ecosystems and the North American carbon cycle. Global Change Biology. 2006;**12**:2092-2102

[54] Coleman K, Jenkinson DS. A Model for the Turnover of Carbon in Soil. Model Description and Windows User Interface. Harpenden, Hertfordshire, UK: Rothampsted Research; 2008

[55] Smith B, Prentice IC, Sykes MT. Representation of vegetation dynamics in the modelling of terrestrial ecosystems: Comparing two contrasting approaches within European climate space. Global Ecology & Biogeography. 2001;**10**(6):621-637

[56] Sitch S, Smith B, Prentice IC, Arneth A, Bondeau A, Cramer W, et al. Evaluation of ecosystem dynamics, plant geography and terrestrial carbon cycling in the LPJ dynamic global vegetation model. Global Change Biology. 2009;**9**(2):161-185

[57] JNCC. Handbook for Phase 1 Habitat Survey—A Technique for Environmental Audit. Peterborough: JNCC; 2010

[58] Gentil E, Christensen TH, Aoustin E. Greenhouse gas accounting and waste management. Waste Management Research. 2009;**27**(8):696-706

[59] Timlett R, Williams I. The ISB model (Infrastructure, Service, Behaviour): A tool for waste practitioners. Waste Management. 2011;**31**(6):1381-1392

[60] Gentil E, Damgaard A, Hauschild MZ, Finnveden G, Eriksson O, Thorneloe S, et al. Models for waste life cycle assessment: Review of technical assumptions. Waste Management. 2010;**30**:2636-2648

[61] Turner D, Williams I, Kemp S, Wright L, Coello J, Mcmurtry E. Towards standardization in GHG quantification and reporting. Carbon Management. 2012;**3**(3):223-225

[62] Lou X, Nair J. The impact of landfilling and composting on greenhouse gas emissions—A review. Bioresource Technology. 2009;**100**(16):3792-3798

[63] Manfredi S, Tonini D, Christensen TH, Scharff H. Landfilling of waste: Accounting of greenhouse gases and global warming contributions. Waste Management Research. 2009;**27**(8):825-836

[64] Walker A. Independent Review of Charging for Household Water and Sewerage Services. London, UK: Department for Environment, Food and Rural Affairs (DEFRA); 2009

[65] Wiedmann T. A review of recent multi-region input-output models used for consumption-based emission and resource accounting. Ecological Economics. 2009;**69**(2):211-222

[66] Miller RE, Blair PD. Input-output Analysis: Foundations and Extensions. Cambridge, UK: Cambridge University Press; 2009

[67] Leontief W. Input Output Economics. Oxford, UK: Oxford University Press; 1986

[68] Office for National Statistics (ONS). Input-output Supply and Use Tables,

2011 Edition [Data File]. London: Copyright TSO; 2011

[69] Office for National Statistics (ONS). UK Environmental Accounts 2012. London: Crown Copyright; 2012

[70] Office of National Statistics (ONS). Inter-Departmental Business Register (IDBR) 2008: Secure Access [Data File]. Crown Copyright; 2009

[71] Office of National Statistics (ONS). Family Spending. 2011th ed. London: Crown Copyright; 2011

[72] Saltelli A, Annoni P. How to avoid a perfunctory sensitivity analysis. Environmental Modeling & Software. 2010;**25**:1508-1517

[73] Lenzen M, Wood R, Weidmann T. Uncertainty analysis for multi-region input-output models—A case study of the UK's carbon footprint. Economic Systems Research. 2010;**22**(1):43-63

[74] United Nations Statistic Division. Environmental Indicators [data File]. United Nations; 2010

[75] Hillman T, Ramaswami A. Greenhouse gas emission footprints and energy use benchmarks for eight U.S. cities. Environmental Science & Technology. 2010;**44**(6):1902-1910

[76] Weber CL, Jaramillo P, Marriott J, Samaras C. Life cycle assessment and grid electricity: What do we know and what can we know? Environmental Science & Technology. 2010;**44**(6):1895-1901

[77] Druckman A, Jackson T. The carbon footprint of UK households 1990-2004: A socio-economically disaggregated, quasi-multi-regional input-output model. Ecological Economics. 2009;**68**(7):2066-2077

[78] Minx JC, Wiedmann T, Wood R, Peters GP, Lenzen M, Owen A, et al. Input-output analysis and carbon footprinting: An overview of applications. Economic Systems Research. 2009;**21**(3):187-216

Chapter 6

Suggestion of Life Cycle Impact Assessment Methodology: Selection Criteria for Environmental Impact Categories

Magali Rejane Rigon, Rafael Zortea,
Carlos Alberto Mendes Moraes
and Regina Célia Espinosa Modolo

Abstract

In life cycle assessment (LCA), environmental impacts are classified according to the methodology used. Several life cycle impact assessment (LCIA) methods are currently used, and the method selected and the particulars thereof may influence the results obtained. This study characterized the main LCIA methods used and the most relevant categories of environmental impact. In total, 87 articles were initially retrieved using relevant keywords. After screening, 11 articles were shown to address the topic of study and were reviewed. The results showed that CML is the most widely used method. The main environmental impact category was global warming potential followed by acidification. Studies using LCA depend on the confirmation of the efficacy of the methods in the effort to represent and assess impacts in different regions of the world.

Keywords: LCA, LCIA methodologies, environmental impacts, global warming potential, acidification

1. Introduction

Life cycle assessment (LCA) is an essential tool in the characterization of environmental risks in the different stages of a product's life cycle [1]. Research on LCA is a source of important information in management and decision-making strategies designed to improve environmental practices and execute technological adjustments or transformations in organizations [2].

With the use of LCA on the increase, the biofuel sector is the object of considerable research publications, followed by energy generation and agriculture [3]. In addition, LCA is the foundation of studies that assess environmental impacts in several production chains such as the steel industry [4], construction [1, 2], steel recycling processes [4], and urban solid waste management [5, 6].

Standardized by the International Organization for Standardization (ISO) as ISO 14.040, the execution of an LCA is divided into four stages, namely the *definition of goal and scope*, *inventory analysis*, *impact assessment,* and *interpretation of results* [7].

In LCA studies, environmental impact is classified according to the methodology used to assess it. The methods used for life cycle impact assessment (LCIA) establish the relationship between each stage of the life cycle inventory and the corresponding environmental impacts [8]. Several LCIA methods based on software and inventory databases have been developed. Notably, the variety and the specific aspects of these methods may affect the end results of LCA [9]. Moreover, it is important to understand the implications of LCA studies from a broader, more inclusive perspective that considers not only the environment but also human health, since important factors may be overlooked if an all-encompassing, holistic approach to environmental impact is not carried out [10]. For this reason, LCA studies require the evaluation of the LCIA method that best characterizes the potential environmental impacts in a given process considering the scope and hypotheses guiding the conduction of study.

The objective of this study was to identify the key LCIA methods used today and characterize the main categories of environmental impact assessed using these methods.

This chapter is structured so as to initially characterize the obstacles and difficulties faced when classifying the environmental impacts central to the conduction of LCA. Next, we carried out a literature review using specific keywords currently used to define the main criteria and the most important categories of environmental impact. Early research already warned of the implications of not including all relevant categories of environmental impact in LCA when comparing the impacts of recycling paper solid waste and incinerating it [11].

2. Methods

Prior to the literature review carried out, we first discussed cases of environmental impact that indicated the importance of a diagnostic evaluation of the selection of all impact categories used in decision-making.

This study was carried out searching the Journals Portal of University Professor Improvement Bureau (CAPES), which includes more than 250 databases of theses, journals, and books. Some of the databases included are SCIELO, Science Direct (Elsevier), and Scopus (Elsevier). The search was carried out in November 2018, and the keywords used were *industrial solid waste* and LCA. Initially, the keywords retrieved 87 publications issued from 1993 to 2018. Subsequently, the articles were screened so as to include only the publications addressing the study topic, namely the use of LCA to investigate solid waste from industrial processes. Some of the articles included were noteworthy literature reviews on the use of LCA [12, 13]. Waste management assessments carried out using other methodologies were excluded from the present report [14, 15]. Similarly, studies that used advanced environmental tools like material flow analysis (MFA) [16, 17], circular economy [18], and industrial symbiosis [19] but did not employ LCA were also disregarded.

The inclusion criteria adopted concern the third stage of LCA studies, namely LCIA. Software, the LCIA method, and the categories of environmental impacts used in these publications were considered. If a study presented assessments of more than one category such as midpoint and endpoint categories like global warming potential and climate change, for example, it was still considered one publication only. The impact categories considered had to be addressed in more than one single publication. However, studies that have produced significant findings were mentioned in the assessment. Later, evidence explaining the selection of environmental impact categories assessed in each study was analyzed.

3. Criteria used to select life cycle impact categories

The use of a LCIA is justified considering the effort to generate a priority matrix that may be used to define the most relevant impact categories in LCA studies. This matrix would be helpful in the characterization of the impact categories that should be considered in each LCA based on specific criteria and score systems.

The criteria to be defined should take into account the importance of each impact in each case studied. Therefore, criteria like *type of process* (that is, the environmental impacts that are more important in a given process and the most affected compartments) and *region* (the scarcest or most susceptible natural resources in a determined area, for example) become important factors to be considered in the definition of the priority matrix. Other important requisites include spatial coverage, duration, reversibility, probability of occurrence, harm to human health, harm to ecosystems, exhaustion of resources, and treatment alternatives [20].

Therefore, based on these criteria, the priority matrix may be helpful in the definition of the most appropriate impact categories to be considered in a given LCA study. It is also important to evaluate the metric that most accurately and realistically represents the categories defined.

For example, the contaminants generated by a given industrial activity are a function of the associated production processes [21]. A metalworking company carries out processes like purification, surface treatments, and quenching. For this reason, such a company would generate by-products like

- foundry sand waste
- cured resin waste
- polymer paint bottoms
- boiler ashes and soot
- quenching salt waste
- galvanizing bath dregs
- metal scrap in general

More specifically, metal processing may generate waste items like metal scrap and sand casting scrap that in turn release phenols, cyanides, mineral oil, and heavy metals. In turn, surface treatments and quenching operations are sources of antimony, arsenic, petroleum ether, benzene, lead, cadmium, chromium, cyanides, copper, mineral oil, nickel, mercury, acids, bases, selenium, and zinc. Investigation on these contamination hazards is an important source of data on the major environmental impact categories in LCA of products manufactured using such processes.

Concerning the situation of a given region in order to assess how its environmental compartments behave and how degraded the region is, the Rio Grande harbor, in southern Brazil, provides a good example. According to Fundação Estadual de Proteção Ambiental Henrique Luiz Roessler (FEPAM), the local environmental authority that records incidents with hazardous materials (http://www.fepam.rs.gov.br/emergencia/rel_acidentes.asp), the accident with the cargo ship Bahamas in 1998 was the first major event in Rio Grande harbor. This ship was transporting concentrated sulfuric acid when it was moored to canal in Rio Grande harbor.

Due to operational problems, sea water leaked into the tanks, diluting the acid in a strongly exothermic process in view of the large cargo (approximately 12,000 t) and considerable volume of water that leaked into the ship [22]. The risk of explosion was significant. The alternative found was to control the release of the sulfuric acid into the canal of the Rio Grande harbor. Fortunately, the canal waters were discharging into the ocean, leading the contaminant to flow away from the harbor area. It should be emphasized that if the waters were flowing into the canal, then the acid would have remained in the harbor area, spreading to Patos Lake and triggering a large-scale environmental disaster due to the ecological susceptibility of the region [22]. From the chemical standpoint, the dilution of sulfuric acid releases high levels of hydroxonium (H_3O^+), sulfate (SO_4^{2-}), and bisulfate (HSO_4^-) ions. Considering that the density of sulfuric acid is higher than that of sea water and that the pH varied from 8 to 3 during the incident (returning to 8 subsequently), it may be hypothesized that solubilized metal ions were stabilized in the aqueous medium by conversion to insoluble sulfate or bisulfate species [22]. This induced the sedimentation of these materials on the floor of the canal.

Three minor accidents took place in the Rio Grande harbor in the past 20 years. Two of these events involved fuel oil and one involved bunker fuel, in 2001 and 2004, respectively. However, since these fluids are poorly soluble and less dense than water, the contamination of the canal floor could be ruled out.

But a lead acetate spill was recorded in the container park of the harbor in July 2001. Different from the accidents with oil, lead acetate was being transported in the solid state. Due to the high solubility in water and high density compared with sea water, the compound posed a high risk of contamination of the canal floor. Though lead acetate is a low-hydrolysis rate organic salt formed from a weak acid, the compound is toxic and the possibility that hydrolysis takes place indicates that Pb^{2+} ions might have reacted with the ions in solution in the waters of the canal of Rio Grande harbor.

In view of that, the record of incidents in the Rio Grande harbor clearly indicates the categories of environmental impacts that are essential to be considered in LCA studies in the region—or in any other harbor zone. These events signal that the activities carried out in a harbor may have high environmental impact hazard.

Considering the prerequisites discussed in previous research [20] and the criteria defined above, which are represented in detail in the accidents described, the present study indicates the need for a priority matrix that addresses these prerequisites. The objective is to provide a decision-making tool in the definition of the main life cycle impact categories to be considered in an LCA.

4. Life cycle impact assessment (LCIA) methods

Of the 87 studies initially retrieved, 11 met the objectives of this review and were appraised. The studies included in this investigation were about LCA of industrial waste like copper tailings [23], management of hazardous industrial waste [24], steel recycling processes [4], solid urban waste management [5, 6], and cement industry [1, 2, 25, 26]. **Table 1** shows the case studies and the environmental impact categories assessed.

It is possible to observe that the LCIA method called CML was the most applied in research, being used in three articles carried out in Turkey, China, and Arabia [1, 5, 6]. The Eco-Indicator 99 method [23] and the Impact2002+ method [2] were also used. Other studies used the software tool developed by IKE Environmental Technology Co. Ltd., the eBalance package, which defines 16 midpoint categories of LCA [26, 27].

Study	Region	Functional unit	Software	LCIA	Impact categories								
					1	2	3	4	5	6	7	8	9
Chen et al. [4]	China	1 kg steel (raw state)	SIMAPRO 7	IPCC 2007	x								
Morris [28]	USA	Solid waste management	BEES 3.0	—	x	x	x						
Song et al. [23]	China	1 t copper	GaBi 4	Eco-Indicator 99	x	x			x		x		x
Al-maaded et al. [6]	Arabia	10 kg plastic waste	GaBi 4	CML 2001	x	x		x		x			
Banar et al. [29]	Turkey	1 t urban solid waste	SIMAPRO 7	CML 2000	x	x	x	x		x		x	
Song et al. [1]	China	1 t Portland cement	NI	CML 2001	x	x	x	x	x	x	x	x	
Yang et al. [2]	China	1 t cement of different resistance values	NI	IMPACT2002+	x	x	x		x				x
Shen et al. [25]	China	Portland cement	Not used	—	x								
Changzai et al. [26]	China	1 t SAC clinker production	IKE Environmental Technology Co. Ltd	eBalance	x	x	x			x			x
Hong et al. [24]	China	1 t of mixed industrial hazardous waste	NI	ReCiPe and USEtox™	x	x	x				x	x	
Wang et al. [16]	China	1 t of coal and 1 MWh power	IKE Environmental Technology Co. Ltd	eBalance	x	x	x						

1: Global warming potential; 2: acidification potential; 3: eutrophication potential; 4: human toxicity potential; 5: ecotoxicity potential; 6: abiotic depletion potential; 7: ozone depletion potential; 8: photochemical oxidation potential; 9: respiratory toxicity potential (inorganic).

Table 1.
Case studies selected and their relationships with environmental impact categories.

The use of the LCIA method is explained based on how a method is applied in LCA studies [2, 23]. However, some studies provide no explanation about the decision concerning the LCIA method selected. The use of a given method based on technical criteria was not reported in any study reviewed.

In LCA studies, environmental impacts are classified into categories based on the methodology used to assess the impact. The selection of categories of environmental impacts to systematically understand the aspects involved in each process is highly important at this stage, since the paucity of information may affect all decision levels.

The results of this review show that the main categories of environmental impacts taken into account in LCIA studies were *global warming potential* and *acidification potential*, which were used in 11 and 9 studies, respectively. *Ecotoxicity* was a category assessed in three studies.

More specifically, the four LCA studies that addressed cement as the only product were carried out in China using different methods. Only one category was assessed in the four studies, namely *global warming potential*. The categories *acidification potential*, *eutrophication potential*, and *ecotoxicity* were evaluated in two studies [1, 2]. The fact that these categories were evaluated does not mean that they were relevant in the respective studies. In the studies that assessed industrial waste and processes, the main categories used were *global warming potential* followed by *acidification potential* and *eutrophication potential.*

Previous research carried out an environmental evaluation of a typical Portland cement production line in China and compared the environmental impacts observed with the best available technologies with effects of the replacement of raw materials and of calcination fuels [1]. The functional unit defined was the production of 1 ton of Portland cement. The data were collected in a company operating in northern China and compiled as a database. The environmental impact categories were assessed using the CML 2001 method. The environmental impacts assessed were normalized. It was possible to observe that the category *global warming potential* is more severe compared with the other categories, followed by *acidification potential* and *photochemical oxidation*. The authors observed that the most efficient way to reduce greenhouse gas emissions in Portland cement production in China includes the study of alternative raw materials and fuels, especially due to the effects of calcination and coal consumption. These results were similar to the findings published in previous research [2], which found that the use of alternative materials like industrial waste and by-products is an efficient way to reduce environmental and economic impact generated in cement production.

The environmental performance of cements produced to yield various resistance levels has been compared [2]. The functional unit chosen was the production of 1 ton of cement. Mean annual production data of cement types were obtained from a research carried out by the China United Cement Corporation. Energy, coal, and shipping data were obtained from the literature. The LCIA method used was Impact 2002+. The environmental impacts were calculated based on midpoint and endpoint categories and were normalized. Based on an LCA, the authors concluded that the cement produced to yield high resistance caused the highest environmental impacts compared with lower resistance cements. The results showed that the categories that most contributed to global environmental impacts are *global warming potential*, *respiratory toxicity potential*, *non-renewable energy consumption*, and *terrestrial acidification/eutrophication*. Therefore, two categories were significant in these studies, namely *global warming potential* and *acidification potential.*

Also, CO_2 emissions by the cement industry in China were quantified using an LCA [25]. Although these authors did not use a specific software, the calculations were carried out based on the necessary equations.

In another study, environmental impacts of the production of sulfoaluminate clinker using industrial solid waste were compared to the results obtained with the conventional method [26]. The results showed that industrial solid waste may significantly reduce the environmental load of the process due to the lower consumption of natural resources and greenhouse gas emissions. The production of sulfoaluminate clinker using industrial waste may reduce the total environmental impact by 38.62% compared with the conventional process.

It has been maintained that most studies about cement production considered only CO_2 emissions and ignored the other environmental impacts [2]. It is observed that this is the case of several LCA studies not only about cement production, but also about other processes. The category *global warming potential* was considered in all studies, which explains the concern of industrial sectors to reduce greenhouse gas emissions.

Another study assessed the generation of energy from coal in China considering the steps of the mining life cycle and the washing and shipping of coal [27]. The authors observed that the main environmental impact category was *smoke and dust*, which is associated with the emission of total suspended particles.

However, it is important to consider all environmental impacts associated with LCA, in view of the relevance of the results of assessments to all decision-making levels. Therefore, it is essential to consider the specific aspects of the regions where a LCA is conducted and identify the relevance of the likely environmental impacts and aspects involved locally.

In addition, decision-makers have to consider a full LCIA, taking into account the associated economic and environmental impacts [4].

The lack of a holistic assessment of environmental impacts was observed in LCA carried out today based on a critical evaluation of LCA studies about concrete [10]. The author reports that LCA studies about concrete published in the literature are based on the use of energy and greenhouse gas emissions, despite the importance of questions like volatile organic compounds, heavy metals, and other toxic emissions involved in the production of concrete components.

5. Final considerations

Based on the rationale presented to determine the selection criteria and the survey carried out about LCA studies on industrial solid waste, it is observed that there is a long way ahead in the definition of a methodology to establish the life cycle environmental impacts that best fit each study in particular. Therefore, it is important to evaluate the methods that include the set of priorities established for the definition of the categories of impact that are of relevance in LCA studies. The priority matrix should include items such as type of activity and overall regional characteristics [20].

6. Conclusions

The objective of this chapter was to evaluate the use of different methods to define the most representative categories of environmental impact in LCA of industrial solid waste. Although initially 87 studies were selected, no study on LCA was carried out using a method that actually helped identify these categories. However, the categories *global warming* prevailed in research, followed by *acidification potential* and *eutrophication potential*.

This chapter also aimed at demonstrating the importance of assessing processes and respective downcycling and upcycling by-products as well as the most frequent pollutants, as in the example of the metalworking organization discussed. The importance of considering the physicochemical characteristics and behavior of compartments like water, air, and soil in the region where an impact occurs is highlighted. It is essential to evaluate the region considering its record of environmental accidents that affect its vulnerability to a given impact category. As opposed to what was observed in this literature review, these peculiarities should not be overlooked, meaning that specific aspects have to be considered in the search for critical points in LCA studies.

In view of that, the present literature review warns of the need to use appropriate LCA methods that consider the factor cited and address spatial area, duration of impact, reversibility, probability of occurring, human health hazards, harm to ecosystems, resource exhaustion, and treatment alternatives. Therefore, research on LCA requires a clearly developed approach to select impact categories that are more relevant in the establishment of environmental critical points, which is one of the objectives of LCA. These considerations form the foundation for a modernized production chain based on sustainable development under research, where LCA is the main tool in decision-making.

Author details

Magali Rejane Rigon[1], Rafael Zortea[2], Carlos Alberto Mendes Moraes[3]* and Regina Célia Espinosa Modolo[3]

1 Polytechnic School, Graduate Program of Civil Engineering of University of Valley of Sinos River (Unisinos/NuCMat), São Leopoldo, Brazil

2 Instituto Federal Sul-Rio-Grandense – IFSUL, Sapucaia do Sul, Brazil

3 Polytechnic School, Graduate Program of Civil Engineering and Mechanical Engineering of University of Valley of Sinos River (Unisinos/NuCMat), São Leopoldo, Brazil

*Address all correspondence to: cmoraes@unisinos.br

References

[1] Song D, Yang J, Chen B, Hayat T, Alsaedi A. Life cycle environmental impact analysis of a typical cement production chain. Applied Energy. 2016;**164**:916-923

[2] Yang D, Fan L, Shi F, Qian L, Wang Y. Comparative study of cement manufacturing with different strength grades using the coupled LCA and partial LCC methods—A case study in China. Resources Conservation and Recycling. 2017;**119**:60-68

[3] Zanghelini GM, Souza HRA Jr, Cherubini E, Kulay L, Soares SR. Análise da evolução dos temas de pesquisa da ACV no Brasil baseada na relação de co-words. Revista Latino-Americana em Avaliação do Ciclo de Vida. 2016. pp. 34-47

[4] Chen B, Yang J, Ouyang Z. Life cycle assessment of internal recycling options of steel slag in Chinese iron and steel industry. Journal of Iron and Steel Research. 2011;**18**(7):33-40

[5] Banar M, Cokaygil Z, Ozkan A. Life cycle assessment of solid waste management options for Eskisehir Turkey. Waste Management. 2009;**20**:54-62

[6] Al-Maaded M, Madi NK, Kahraman R, Hodzic A, Ozerkan NG. An overview of solid waste management and plastic recycling in Qatar. Journal of Polymers and the Environment. 2012;**20**:186-194

[7] Associação Brasileira de Normas Técnicas (ABNT). NBR ISO 14040: Gestão Ambiental: Avaliação do ciclo de vida: Princípios e estrutura. Rio de Janeiro: ABNT; 2009

[8] Jolliet O, Margni M, Charles R, Rosenbaum RK. IMPACT 2002+ A new life cycle assessment methodology. The International Journal of Life Cycle Assessment. 2003;**8**(6):324-330

[9] Mendes NC. Métodos e modelos de caracterização para a Avaliação de Impacto do Ciclo de Vida: análise e subsídios para a aplicação no Brasil. Dissertation. São Paulo: Universidade de São Paulo; ano

[10] Gursel AP, Maryman H, Ostertag C. A life-cycle approach to environmental mechanical and durability properties of "green" concrete mixes with rice husk ash. Journal of Cleaner Production. 2016;**112**:823-836

[11] Finnveden G, Ekvall T. Life-cycle assessment as a decision-support tool—The case of recycling versus incineration of paper. Resources Conservation and Recycling. 1998;**24**:235-256

[12] Othman SN, Noor ZZ, Abba AH, Yusuf RO. Review on life cycle assessment of integrated solid waste management in some Asian countries. Journal of Cleaner Production. 2013;**41**:251-262. DOI: 10.1016/j.jclepro.2012.09.043

[13] Cleary J. Life cycle assessments of municipal solid waste management systems: A comparative analysis of selected peer-reviewed literature. Environment International. 2009;**35**:1256-1266. DOI: 10.1016/j.envint.2009.07.009

[14] Usapein P, Chavalparit O. Development of sustainable waste management toward zero landfill waste for the petrochemical industry in Thailand using a comprehensive 3R methodology: A case study. Waste Management & Research. 2014;**32**:509-518. DOI: 10.1177/0734242X14533604

[15] Coelho LMG, Lange LC, Coleho HMG. Multi-criteria decision making to support waste management: A critical review of current practices and methods. Waste Management

& Research. 2017;**35**:3-28. DOI: 10.1177/0734242X16664024

[16] Wang W, Jiang D, Chen D, Chen Z, Zhou W, Zhu B. A material flow analysis (MFA)-based potential analysis of ecoefficiency indicators of China's cement and cement-based materials Industry. Journal of Cleaner Production. 2016;**112**:787-796. DOI: 10.1016/j.jclepro.2015.06.103

[17] Vahidi H, Hoveidi H, Khoie JK, Nematollahi H, Heydari R. Analyzing material flow in Alborz industrial estate Ghazvin Iran. Journal of Material Cycles Waste Management. 2018;**20**:450-460. DOI: 10.1007/s10163-017-0601-9

[18] Veleva V, Bodkin G. Corporate-entrepreneur collaborations to advance a circular economy. Journal of Cleaner Production. 2018;**188**:20-37. DOI: 10.1016/j.jclepro.2018.03.196

[19] Sun L, Li H, Dong L, Fang K, Ren JH, Geng Y, et al. Eco-benefits assessment on urban industrial symbiosis based on material flows analysis and emergy evaluation approach: A case of Liuzhou city China. Resources Conservation and Recycling. 2017;**119**:78-88. DOI: 10.1016/j.resconrec.2016.06.007

[20] Zanghelini GM. Ponderação de categorias de impacto ambiental através de análise de decisão multicritério [thesis]. Florianópolis: Universidade Federal de Santa Catarina; 2018

[21] Schianetz B. Passivos Ambientais–Levantamento Histórico Avaliação da Periculosidade e Ações de Recuperação. SENAI: Curitiba; 1999. 200p

[22] Pereira RS, Niencheski LFH, Vitola MA, Pinto WT. Simulação computacional do acidente com o navio tanque Bahamas no porto de Rio Grande. In: Seminário e Workshop em Engenharia Oceânica. 2004; Rio Grande. Ed. da FURG 2004. p. 17

[23] Song X, Yan J, Lu B, Li B. Exploring the life cycle management of industrial solid waste in the case of copper slag. Waste Management & Research. 2013;**31**(6):625-633

[24] Hong J, Han X, Chen Y, Wang M, Ye K, Qi Cm Ki X. Life cycle environmental assessment of industrial hazardous waste incineration and landfilling in China. Journal of Life Cycle Assessment. 2017;**22**:1054-1064

[25] Shen W, Cao L, Li Q, Zhang W. Quantifying CO_2 emissions from China's cement industry. Renewable and Sustainable Energy. 2015;**50**:1004-1012

[26] Changzai R. Comparative life cycle assessment of sulfoaluminate clinker production derived from industrial solid wastes and conventional raw materials. Journal of Cleaner Production. 2017;**167**:1314-1324

[27] Wang J, Wang R, Zhu Y, Li J. Life cycle assessment and environmental cost accounting of coal-fired power generation in China. Energy Policy. 2018;**115**:374-384

[28] Morris J. Comparative LCAs form curbside recycling versus either landfill or incineration with energy recovey. InLCA case Studies. 2005:**10**:273-284

[29] Banar M, Cokaygil Z, Ozkan A. Life cycle assessment of solid waste management options for Eskisehir, Turkey. ScienceDirect. 2008;**29**:54-62. DOI: 10.1016/j.wasman.2007.12.006